T0303911

Symbolic Computation and
Automated Reasoning

pp. ii

blank

Symbolic Computation and Automated Reasoning

The CALCULEMUS-2000 Symposium

Edited by
Manfred Kerber
Michael Kohlhase

CRC Press
Taylor & Francis Group
Boca Raton London New York

CRC Press is an imprint of the
Taylor & Francis Group, an **informa** business

CRC Press
Taylor & Francis Group
6000 Broken Sound Parkway NW, Suite 300
Boca Raton, FL 33487-2742

Preface

Both deduction systems and computer algebra systems are receiving growing attention from industry and academia. On the one hand, mathematical software systems have been commercially very successful. Their use is now wide-spread in industry, education, and scientific contexts. On the other hand, the use of formal methods in hardware and software development has made deduction systems indispensable not least because of the complexity and sheer size of the reasoning tasks involved. As many application domains fall outside the scope of existing deduction systems and computer algebra systems, there is still need for improvement and in particular need for the integration of computer algebra and deduction systems.

The CALCULEMUS symposia are intended for researchers and developers interested in combining the reasoning capabilities of deduction systems and the computational power of computer algebra systems. We want to form a bridge between the deduction and computer algebra communities by providing a discussion forum for novel work on the integration of such systems, as well as work on mathematical software systems that was motivated by or can contribute to the goal of integrating deduction and symbolic computation systems.

A core group from the CALCULEMUS interest group has applied for and received a networking grant in the framework of the European Union research programme under the IHP (Improving Human Potential) programme. The CALCULEMUS symposia will serve as a primary discussion forum for this research network.

Information on CALCULEMUS can be found online as

http://www.calculemus.net/

Topics of interest for 8th Symposium on the Integration of Symbolic Computation and Mechanized Reasoning, 6–7 August 2000 in St Andrews, Scotland, include all aspects related to the combination of deduction systems and computer algebra systems. We have explicitly encouraged submissions of results from applications and case studies where such integration results are particularly important.

We would like to thank the members of the programme committee and all the referees for their important work in selecting the submitted papers. We had 29 submissions, out of which we selected 14 as regular papers, one as system description, and eight as posters. The submitted posters were complemented by posters of participating nodes about their work.

All papers and posters were presented at the symposium. For the proceedings in this book, they were revised to take into account the discussion there. The symposium was collocated with ISSAC 2000 (International Symposium on Symbolic and Algebraic Computation), with which we joined one of the two invited presentations.

Manfred Kerber, Michael Kohlhase
Co-chairs

Programme Committee

Alessandro Armando	U. Genova
Michael Beeson	San Jose State U.
Manuel Bronstein	INRIA Sophia Antipolis
Bruno Buchberger	RISC, Linz
Jaques Calmet	U. Karlsruhe
Olga Caprotti	TU. Eindhoven
Edmund Clarke	CMU
Fausto Giunchiglia	IRST
Thérèse Hardin	Paris VI
John Harrison	Intel Corp.
Tudor Jebelean	RISC, Linz
Deepak Kapur	U. New Mexico, Albuquerque
Manfred Kerber	U. Birmingham (Co-chair)
Helene Kirchner	Nancy LORIA/INRIA
Michael Kohlhase	U. Saarbrücken (Co-chair)
Steve Linton	St. Andrews U. (Local Org.)
Ursula Martin	St. Andrews U.
Julian Richardson	Heriot Watt U., Edinburgh
Jörg Siekmann	U. Saarbrücken
Carolyn Talcott	Stanford U.
Andrzej Trybulec	U. Białystok

Reviewers (other than PC members)

Serge Autexier	Clemens Ballarin
Grzegorz Bancerek	Simon Colton
Ewen Denney	Dominique Duval
Roy Dyckhoff	Paul Libbrecht
Heiko Mantel	Martijn Oostdijk
Martin Pollet	Loic Pottier
Silvio Ranise	Renaud Rioboo
Freek Wiedijk	Daniele Zini

Previous meetings:

16 – 19 March 1996, Rome, Italy
July 1996, Workshop on IMACS
18-20 November 1996, Dagstuhl, Germany
28-30 April 1997, IRST, Italy
24 - 26 September 1997, Edinburgh, Scotland
13-15 July 1998, Eindhoven, The Netherlands
5 July 1999, Trento, Italy

Contents

Part I

Regular Contributions

Definite Integration of Parametric Rational Functions: Applying a DITLU

Andrew A. Adams*

Abstract. *In [2] we presented a Definite Integral Table Lookup (the DITLU) for parametric functions, including a minimal prototype implementation demonstrating its capabilities. In this paper we present a possible application of a DITLU, which would extend its utility for a modest investment of effort. The naive algorithm for indefinite integration of rational functions (see e.g. [12, §2.10]) can be implemented for parametric rational functions. This involves splitting the rational function integrand using partial fractions. The resulting integrands all fall within a limited class which may be covered in a DITLU by a very small number of table entries. Extensions of this idea to less naive integration algorithms, and the number of table entries required to implement them, are also considered.*

1 Introduction

Definite integration is acknowledged [14] to be a tricky problem for Computer Algebra Systems [CAS] when a symbolic answer is required or when the original query involves parameters, i.e. in those cases where the well-developed numeric methods are inapplicable. In [2] and elsewhere we presented a Definite Integral Table Lookup (the DITLU) for calculating definite integrals involving parameters in the integrand and the limits. We implemented a prototype version of the table to perform proof of concept, and described the extensions necessary for a production version of the system to be included in a CAS. The DITLU uses automated theorem proving (ATP) technology to winnow inconsistent cases from the multiplicity that occur in examples with a number of parameters. Our previous papers

*This work is supported by the UK EPSRC: grant GR/L48256.

3

([1, 2, 3]) also described how the contents of the DITLU might be formally verified correct using machine assisted theorem proving (the difficulty of the proofs is beyond current automatic techniques). While finalising the form of the prototype it occurred to us that some simple extensions of the table would allow it to solve far more cases than those actually included in the table. In this paper we present an example of this: the extensions to the system and the table entries required for the DITLU to be instrumental in the correct calculation of any definite integral of a parametric rational function.

First we begin with a review of our DITLU system, in Section 2. Then in Section 3 we present the naive algorithm for integrating rational functions with parameters using the DITLU. In Section 4 we consider other indefinite integration algorithms and how their benefits could be included in our methods. Our conclusions are presented in Section 5. In the appendix we include the table entries that would need to be added to the DITLU for the naive algorithm to be used.

2   Review of the DITLU

Definite integration in the presence of parameters is a tricky problem, and one for which CAS often return incorrect, or partially correct, answers. Much of this problem is due to design issues, mainly historical, which have produced programs very good at performing algebraic calculation, but poor at analytic calculation. One of the primary problems for CAS designers when attempting to answer analytic queries is that analysis is much concerned with ranges of definition, points of undefinedness and actual values of functions, whereas algebra is a much more abstract topic, in which any function satisfying $D(f) = f$, where D is a differential operator, is regarded as *an* exponential function. Analysis, on the other hand is concerned with the difference, as much as the similarities, between e^x and $e^{(x+1)}$. While CAS are designed to calculate algebraic quantities they are not good at keeping track of the logical side conditions of analysis. There have been many attempts to develop mathematical software which retains a high level of logical consistency, but mostly these have ended up quite well separated from the mainstream user of mathematical software such as Maple or Mathematica. Our approach was to consider the shortcomings of existing CAS and consider how ATP technology could be brought to bear upon these problems. One result of our efforts is the DITLU. The DITLU consists of a table of parametric definite integrals. A user may submit a query to the table with complicated constraints on the side conditions. It is our opinion that useful parametric queries are often highly constrained, and that the inability of CAS to cope well with the analytic ramifications of such constraints is one of their weaknesses. For example, a user may be interested in:

$$\int_0^b \frac{1}{x-1}\, dx \qquad (b > 0) \wedge (b \neq 1) \tag{1}$$

while allowing the use of the Cauchy Principal Value method, which allows positive and negative areas on either side of a pole to cancel each other out. The CAS **axiom** has no ability to perform definite integration through poles using the Cauchy Principal Value method. Design flaws and the inability to cope well with side conditions on parameters lead Maple, Mathematica and Matlab to all return $\ln(b - 1) - i\pi$, which is correct for $b < 1$, albeit only via complex arithmetic, but incorrect for $b > 1$, giving a complex answer. The prototype DITLU can produce the correct answer to this query, since it contains the table entry shown in Figure 1.

To accommodate the limitations of pattern matching in the prototype we submit our query thus:

$$\int_0^k \frac{1}{1.x-1}\, dx \qquad (k > 0) \wedge (k \neq 1) \tag{2}$$

and the system then matches it against the entry shown in Figure 1, with table variables matched thus:

$$q \to 1; \qquad r \to -1; \qquad b \to 0; \qquad c \to k \tag{3}$$

The prototype DITLU then calls the theorem prover PVS to attempt to prove eight conjectures, corresponding to the eight parts of the answer in Figure 1. Each conjecture represents an attempt to prove that a part of the answer is unnecessary for the query submitted because the constraints on the query cannot be satisfied together with the constraints on that part of the answer. Here is an example of a conjecture that is proved by PVS, allowing the system to remove the first line in Figure 1 from the answer returned:

$$\neg \exists k : \mathbb{R}.(k = 0) \wedge ((k > 0) \wedge (k \neq 1)) \tag{4}$$

The first part of this ($k = 0$) is the constraints from the table entry ($b = c$), dereferenced according to the variable assignments in (3), the rest are the query constraints. Similarly the table will be able to prove that lines 1, 2,

$$\int_b^c \frac{1}{qx+r}\,dx \;=\;$$

	Result	*Constraints*
1	0	$(b = c)$
2	Undefined	$(b \neq c) \wedge (q \neq 0) \wedge$ $\left(\left(b = \dfrac{-r}{q} \right) \vee \left(c = \dfrac{-r}{q} \right) \right)$
3	$\ln|qc+r| - \ln|qb+r|$	$(b \neq c) \wedge (q \neq 0) \wedge$ $\left(b < \dfrac{-r}{q} \right) \wedge \left(c < \dfrac{-r}{q} \right)$
4	$\ln|qc+r| - \ln|qb+r|$	$(b \neq c) \wedge (q \neq 0) \wedge$ $\left(b > \dfrac{-r}{q} \right) \wedge \left(c > \dfrac{-r}{q} \right)$
5	$\ln|qc+r| - \ln|qb+r|$ (CPV)	$(b \neq c) \wedge (q \neq 0) \wedge$ $\left(b < \dfrac{-r}{q} \right) \wedge \left(c > \dfrac{-r}{q} \right)$
6	$\ln|qc+r| - \ln|qb+r|$ (CPV)	$(b \neq c) \wedge (q \neq 0) \wedge$ $\left(b > \dfrac{-r}{q} \right) \wedge \left(c < \dfrac{-r}{q} \right)$
7	$\dfrac{c-b}{r}$	$(b \neq c) \wedge (r \neq 0) \wedge (q = 0)$
8	Undefined	$(b \neq c) \wedge (r = 0) \wedge (q = 0)$

Figure 1. A DITLU Entry.

4, 6, 7 and 8 have unsatisfiable constraints for this query and remove them. The other lines give rise to false, and therefore unprovable, conjectures such as:

$$\neg\exists k : \mathbb{R}.((0 \neq k) \wedge (1 \neq 0) \wedge (0 < 1) \wedge (k > 1)) \wedge \tag{5}$$
$$((k > 0) \wedge (k \neq 1))$$

So the answer the DITLU returns is:

$$\ln|1 \times k - 1| - \ln|1 \times 0 - 1| \qquad \frac{(0 \neq 1) \wedge (1 \neq 0) \wedge}{(0 < 1) \wedge (k < 1)} \tag{6}$$

$$\ln|1 \times k - 1| - \ln|1 \times 0 - 1| \qquad \frac{(0 \neq k) \wedge (1 \neq 0) \wedge}{(0 < 1) \wedge (k > 1)} \tag{7}$$
$$(\text{CPV})$$

One might ask why a theorem prover is used for this constraint checking, rather than a decision procedure for real arithmetic or a constraint solver. The answer to this is that at present, the prototype implementation may indeed be no more successful than these techniques. However, the use of a theorem prover provides scope for extensions impossible with other approaches. For instance, the prototype implementation makes use of the equational aspects of the deduction engine to support the matching algorithm, in a similar way to the MFD2 system [6]. Existing constraint satisfaction systems all work with floating point representations and therefore use, for instance, approximations to irrational numbers such as π. Decision procedures for the real numbers are only available for very limited problem sets and extension of the constraints to include transcendental functions of the parameters quickly puts the constraint satisfaction problem outside the application are for such systems. We are currently exploring ideas for solving such problems automatically with PVS, using ideas such as those proposed by Sterling et al in [13].

2.1 Details of the Prototype

The prototype implementation is written in Allgero Common Lisp and comprises approximately 5000 lines of code. We have also extended the existing PVS real analysis library by Dutertre [8] to include some transcendental functions (exp, log and the trig and inverse trig functions) [11].

We are currently considering the existing automation available in PVS and its utility in proving constraints involving transcendental functions.

The interface between PVS and the prototype is crude, consisting of output printed by the lisp session which may then be copied and pasted into a PVS session. Since PVS is also implemented in Allegro Common Lisp, we are investigating ways to link the two systems more closely.

3   Naive Algorithm for Integrating Rational Functions

In this section we present the method for using a small number of simple integrals as DITLU table entries to calculate definite integrals for all rational functions of one variable over the reals. The algorithm is based on the naive method for calculating indefinite integrals for rational functions.

We are considering quotients of polynomials which take real values of a single variable. We are only considering parameters which appear in the coefficients of the polynomials, not in the exponents. In order to use the DITLU to integrate such functions it might appear that an infinite number of entries would be required. In fact even allowing the coefficients of the polynomials to contain transcendental functions in the parameters, all rational functions can be decomposed to use one of four DITLU entries. Much pre-processing of the original rational function integrand is required however, and the complexity of the pre-processing appears to be very high (see Equation (8) for an example of the pre-processing required). Nevertheless we feel this is an interesting possible application/extension of the DITLU.

The following algorithm is described in [12, §2.10] for indefinite integration, and the table entries are developed from the indefinite integrals shown in that section.

From an original query of:

$$\int_t^u \frac{P_1(x)}{P_2(x)} \ dx, C$$

where C is a set of constraints on the parameters occurring in t, u, P_1 and P_2, we first reduce the integral to a sum of integrals of the form:

$$\int_t^u P_3(x) \ dx + \int_t^u \frac{P_4(x)}{P_2(x)} \ dx, C'$$

where $\deg(P_4) < \deg(P_2)$. The first half of the sum is always a well-defined integral that current algorithms can deal with satisfactorily. Errors in the output from CAS implementations of these algorithms are caused by bugs elsewhere within the systems and not our concern. So, we will focus on the problem of integrating rational functions where the degree of the denominator is strictly greater than the degree of the numerator. The main problem with definite integration derives from discontinuities in the integrand. For the rational functions such discontinuities only occur at easily defined places: at the roots of the denominator polynomial. So, we must calculate the roots of P_2. This is a computationally hard problem in the parametric case, but not a difficult algorithm. ATP technology might well be of use here in ruling out cases with inconsistent constraints on the values of the parameters. P_2 may be decomposed thus:

$$P_2 = Q_1^{n_1} \ldots Q_k^{n_k} t^{m_1} \ldots L_j^{m_j}$$

where the Q_i are all quadratics with no real roots and the L_i are linear and hence have exactly one real root. For example, we may decompose:

$$x^3 + cx$$

to

$$x(x + \sqrt{-c})(x - \sqrt{-c}) \qquad \text{if } c < 0$$

$$(x)^3 \qquad \text{if } c = 0$$

$$x(x^2 + c) \qquad \text{if } c > 0.$$

From the factorisation of P_2 we may use partial fractions to decompose:

$$\frac{P_4}{P_2} = \sum_{i=1}^{k} \left(\frac{R_i}{Q_i^{n_i}} \right) + \sum_{i=1}^{j} \left(\frac{S_i}{L_i^{m_i}} \right)$$

where $\deg(R_i) < 2n_i$ and $\deg(S_i) < m_i$. If $\deg(R_i) > 1$ or $\deg(S_i) > 0$, we may decompose them thus:

$$\frac{R}{Q^n} = \frac{a_1 x + b_1}{Q} + \frac{a_2 x + b_2}{Q^2} + \ldots + \frac{a_n x + b_n}{Q^n}$$

$$\frac{S}{L^m} = \frac{c_1}{L} + \frac{c_2}{L^2} + \ldots + \frac{c_m}{L^m}.$$

A more concrete example illustrates the factorisation more clearly:

$$\frac{x}{(x+1)^3} = \frac{0}{(x+1)} + \frac{1}{(x+1)^2} + \frac{(-1)}{(x+1)^3}.$$

So, if we are calculating:

$$\int_t^u \frac{1}{x^3 + cx} \, dx \qquad (8)$$

we decompose this into its factors and partial fractions thus:

$$\frac{1}{c}\int_t^u \frac{1}{(x)^1}\,dx - \frac{1}{2c}\int_t^u \frac{1}{(x+\sqrt{-c})^1}\,dx-$$
$$\frac{1}{2c}\int_t^u \frac{1}{(x-\sqrt{-c})^1}\,dx \quad \text{if } c < 0$$

$$\int_t^u \frac{1}{(x)^3}\,dx \qquad\qquad \text{if } c = 0$$

$$\frac{1}{c}\int_t^u \frac{1}{(x)^1}\,dx - \frac{1}{c}\int_t^u \frac{x}{(x^2+c)^1}\,dx \qquad \text{if } c > 0.$$

Note that extraneous constant factors have been pushed outside the integrals.

So, if we had DITLU entries covering the following integrals, we can deal with all rational integrals in full safety.

$$\int_t^u \frac{m}{(x+c)^i}\,dx \qquad i \in \mathbf{Z}^+$$
$$\int_t^u \frac{mx+n}{(x^2+bx+c)^i}\,dx \quad i \in \mathbf{Z}^+ \wedge b^2 < 4c$$

The general form of these integrals (see Figures 2 and 3) as produced from the indefinite integrals [12, §2.103,{1–5}] by applying the Fundamental Theorem of Calculus, includes recursive entries (both entries for "quadratic" queries are recursive). Our prototype does not include a method of entering recursive integrals into the table, but there is no apparent reason why such a system could not be included in a production version of the DITLU. Note that answers for these entries which would not be necessary for this algorithm have been entered as "Unknown" for the sake of brevity.

We must note that there is a high computational complexity associated with the factorisation step involved in this algorithm. Full factorisation of a parametric rational function is inherently an expensive operation. The ATP technology used to support the DITLU should be equally as useful, however, in supporting this parametric factorisation, by helping to identify unsatisfiable sets of constraints on branches of the factorisation. It is our contention that any parametric query that is useful will be highly constrained. It also seems reasonable that highly constrained queries will factorise without too many case splits on parameter values (randomly this may not be the case but for problems of interest we feel this is a reasonable assumption). Provided we have a factorisation algorithm that copes with constraints on the values of the parameters, we feel that the naive algorithm is a reasonable approach in itself. This naive algorithm is not used in CAS because of its high computational complexity and also because of the introduction of unnecessary algebraic numbers into the answer. We therefore consider the possible utility of using the DITLU in a similar fashion with the more advanced algorithms used in computing indefinite integrals of rational functions.

4 Extensions to the Algorithm

In this section we consider the common variant methods used in calculating indefinite integrals of rational functions, their application to the problem of definite integration, and how this affects our proposal. We begin with a brief discussion of other methods in Subsection 4.1, and then discuss whether those gains carry over to the definite integration case in Subsection 4.2. We discuss in Section 5 how we might change our algorithm above to make use of the efficiency gains of these other methods while retaining confidence in the answer.

4.1 Other Methods

We begin with a parametric definite integral of a rational function:

$$\int_a^b \frac{N}{Q} \, dx. \tag{9}$$

As mentioned above, we are only interested in cases where $\deg(N) < \deg(Q)$, since the definite integration of polynomials themselves is a very simple calculation requiring only care in the programming to be robust and complete, even in the presence of parameters.

The Ostrogradskiy-Hermite Method

The Ostrogradskiy-Hermite Method [12, §2.104] (also called simply Hermite's Method in [10, §11.3]) simplifies the indefinite integral of a rational function thus:

$$\int \frac{N}{Q} \, dx = \frac{P}{R} + \int \frac{L}{M} \, dx$$

where $\deg L < \deg M$ and M is monic and square-free. The rational function $\frac{P}{R}$ is referred to as the rational part of the integral. The indefinite integral remaining is called the logarithmic part of the integral, because it requires answers involving logarithmic extensions.

The core of Hermite's method is a square-free factorisation of the original denominator polynomial, combined with some use of integration by parts. Therefore, using Hermite's method on a rational function involving parameters would appear to be no more difficult than factoring the original parametric polynomial Q. However, since we are concerned with definite integrals we must remember:

$$\int_a^b \frac{N}{Q}\, dx = \left[\frac{P}{R}\right]_a^b + \int_a^b \frac{L}{M}\, dx$$

so that care must be taken in substituting the limits into the rational part. In fact, this substitution is as tricky as any other definite integration of a rational function in that we must be aware of any roots of R in $[a, b]$.

Considering this limit substitution we realise that the solution to this problem requires only a simple extension of the DITLU. An identical matching algorithm and constraint checking system can be applied to the problem of limit substitution, so all we need to apply the DITLU to this problem is to limit substitution is a flag distinguishing between integrals and limit substitutions. Two tables, one of integrals and one of limit substitutions could then use the same code.

Horowitz' Method

Horowitz' method is a variation of Hermite's method for calculating the rational part of the indefinite integral. Horowitz' method would appear no more difficult to apply in the parametric case than Hermite's method, and so for our purposes the result is the same.

The Rothstein-Trager Method

We now turn our attention to the methods for calculating the logarithmic part of the integral. The aim of this method, and others such as the Lazard-Rioboo-Trager improvement, is to avoid performing calculations in any larger an extension field than is absolutely necessary. It is fairly common for the naive method described in Section 3 to produce answers involving spurious algebraic extensions such as $\sqrt{a}$, $\sqrt{a\sqrt{b}}$ etc. which may be simplified out into expressions simply involving a and b. For example:

$$p > 0 \qquad \frac{x^3}{x^4 + p^2} = \frac{x^3}{(x^2 - \sqrt{2p}x + p)(x^2 + \sqrt{2p}x + p)}$$

and so the answer provided by the naive method would involve $\sqrt{2p}$, whereas since

$$p > 0 \Rightarrow x^4 + p^2 > 0 \text{ and } \frac{d}{dx}(x^4 + p^2) = 4x^3,$$

the simple result of a definite integral is:

$$p > 0 \qquad \int_a^b \frac{4x^3}{x^4 + p^2}\,dx = \ln(b^4 + p^2) - \ln(a^4 + p^2)$$

which does not involve algebraic extensions.

The extension of Rothstein-Trager and its improvements to include parametric rational functions is not something that has been done in any existing CAS, although Bronstein's group are currently working on implementing various parametric algorithms properly, as opposed to the ad hoc implementations currently in use, as part of their $\Sigma^I T$ library [4].

4.2 Discussion of Other Methods

In the absence of implementations of the more advanced methods such as Rothstein-Trager and Horowitz for parametric rational functions, it is difficult to judge how much benefit they bring to the problem in this case. Such methods were at least partly developed because computation in algebraic extension fields was expensive. This is almost certainly no longer the case. There is also something to be said for implementing simple algorithms when one is at least as interested in the validity of the output as in the time taken to compute that output.

5 Conclusions

In order to use the above methods instead of a full factorisation one would need to extend the DITLU in two ways. More table entries would be needed than those shown in the appendix. Our original algorithm required only four table entries (fewer with a more sophisticated DITLU than our prototype) to cover any rational function. We need only consider one table entry for each degree of the denominator, since

$$\frac{a_n x^n + \cdots + a_1 x + a_0}{b_{n+1} x^{n+1} + \ldots + b_1 x + b_0}$$

matches

$$\frac{c_k x^k + \cdots + c_1 x + c_0}{d_{n+1} x^{n+1} + \ldots + d_1 x + d_0}$$

for any $k \leq n$, with $a_i = 0$ for $i > k$.

We may assume that for any well-constrained problem there will be a reasonable level of factorisation, and so entries up to denominators of degree n will be sufficient to allow full answers to be calculated for a large number of queries up to degree n^2, and a decreasing number of queries of higher degree.

This leaves the problem of computing the limit substitution for the "rational part" of the indefinite integral. At first this may appear a relatively difficult problem but it is also soluble using a variant of a DITLU. There is almost no difference between table entries comprising integrals of rational functions and limit substitutions into rational functions. Only the table compilation is different. Table operation is identical provided we add a distinguishing marker to differentiate between integrals and limit substitutions.

So, whether using the simple, yet computationally expensive, naive method, or the more complicated and more efficient methods, the problems of using the indefinite integral to compute a definite integral may be bypassed by astute use of a DITLU. A limited number of entries may be used for computing a much larger number of queries, including those of higher degree.

References

[1] A. A. Adams, H. Gottliebsen, S. A. Linton, and U. Martin. A Verifiable Symbolic Definite Integral Table Look-Up. Tech. Rep. CS/99/3, University of St Andrews, 1999.

[2] A. A. Adams, H. Gottliebsen, S. A. Linton, and U. Martin. Automated theorem proving in support of computer algebra: symbolic definite integration as a case study. In Dooley [7], pp. 253–260.

[3] A. A. Adams, H. Gottliebsen, S. A. Linton, and U. Martin. VSDITLU: a verifiable symbolic definite integral table look-up. In Ganzinger [9], pp. 112–126.

[4] M. Bronstein. SUM-IT: A strongly-typed embeddable computer algebra library. In Calmet and Limongelli [5].

[5] J. Calmet, and C. Limongelli, Eds. *Design and Implementation of Symbolic Computation Systems, International Symposium, DISCO '96* (1996), Springer-Verlag LNCS 1128.

[6] S. Dalmas, M. Gaëtano, and C. Huchet. A Deductive Database for Mathematical Formulas. In Calmet and Limongelli [5].

[7] S. Dooley, Ed. *Proceedings of the 1999 International Symposium on Symbolic and Algebraic Computation* (1999), ACM Press.

[8] B. Dutertre. Elements of Mathematical Analysis in PVS. In von Wright et al. [16], pp. 141–156.

[9] H. Ganzinger, Ed. *Automated Deduction — CADE-16* (1999), Springer-Verlag LNAI 1632.

[10] K. O. Geddes, S. R. Czapor, and G. Labahn. *Algorithms for Computer Algebra.* Kluwer, 1992.

[11] H. Gottliebsen. Transcendental Functions and Continuity Checking in PVS. In TPHOLS00 [15]. Submitted.

[12] I. S. Gradshteyn, and I. M. Ryzhik. *Table of Integrals, Series and Products.* Academic Press, 1965.

[13] L. Sterling, A. Bundy, L. Byrd, R. O'Keefe, and B. Silver. Solving symbolic equations with PRESS. *J. Symbolic Comput. 7*, 1 (1989), 71–84.

[14] D. Stoutemyer. Crimes and misdemeanours in the computer algebra trade. *Notices of the AMS 38* (1991), 779–785.

[15] *Theorem Proving in Higher Order Logics* (2000).

[16] J. von Wright, J. Grundy, and J. Harrison, Eds. *Theorem Proving in Higher Order Logics: 9th International Conference* (1996), Springer-Verlag LNCS 1125.

Appendix: Table Entries for the Naive Algorithm

$$i \in \mathbf{Z} \quad \int_t^u \frac{dx}{(x+c)^{2i}} =$$

Result	Constraints
Result	*Constraints*
0	$t = u$
unknown	$(t \neq u) \wedge (i < 1)$
undefined	$(t \neq u) \wedge (i > 0) \wedge$ $((t = -c) \vee (u = -c))$
undefined	$(t \neq u) \wedge (i > 0) \wedge$ $(((t < -c) \wedge (u > -c)) \vee$ $((t > -c) \wedge (u < -c)))$
$\dfrac{1}{(2i-1)} \left(\dfrac{1}{(t+c)^{2i-1}} - \dfrac{1}{(u+c)^{2i-1}} \right)$	$(t \neq u) \wedge (i > 0) \wedge$ $(((t < -c) \wedge (u < -c)) \vee$ $((t > -c) \wedge (u > -c)))$

$$i \in \mathbf{Z} \quad \int_t^u \frac{dx}{(x+c)^{(2i+1)}} =$$

Result	Constraints
0	$t = u$
unknown	$(t \neq u) \wedge (i < 0)$
undefined	$(t \neq u) \wedge (i \geq 0) \wedge$ $((t = -c) \vee (u = -c))$
$\ln\left(\dfrac{|u+c|}{|t+c|} \right)$	$(t \neq u) \wedge (i = 0) \wedge$ $(t \neq -c) \wedge (u \neq -c)$
$\dfrac{1}{2i} \left(\dfrac{1}{(t+c)^{2i}} - \dfrac{1}{(u+c)^{2i}} \right)$	$(t \neq u) \wedge (i > 0) \wedge$ $(t \neq -c) \wedge (u \neq -c)$

Figure 2. Entries for "Linear" Queries.

$$i \in \mathbf{Z} \quad \int_t^u \frac{(mx+n)}{(x^2+bx+c)^i}\, dx =$$

0	$t = u$
unknown	$(t \neq u) \wedge$ $((i < 1) \vee (b^2 \geq 4c))$
$\dfrac{m}{2}\ln\left(\dfrac{(u^2+bu+c)}{(t^2+bt+c)}\right) +$ $\dfrac{(2n-mb)}{\sqrt{4c-b^2}}\left(\tan^{-1}\left(\dfrac{(2u+b)}{\sqrt{4c-b^2}}\right) - \tan^{-1}\left(\dfrac{(2t+b)}{\sqrt{4c-b^2}}\right)\right)$	$(t \neq u) \wedge$ $(i = 1) \wedge (b^2 < 4c)$
$\dfrac{nb - 2mc + (2n-mb)u}{(i-1)(4c-b^2)(u^2+bu+c)} -$ $\dfrac{nb - 2mc + (2n-mb)t}{(i-1)(4c-b^2)(t^2+bt+c)} +$ $\dfrac{(2i-3)(2n-mb)}{(i-1)(4c-b^2)}\displaystyle\int_t^u \frac{1}{(x^2+bx+c)^{i-1}}\, dx$	$(t \neq u) \wedge$ $(i > 1) \wedge (b^2 < 4c)$

$$i \in \mathbf{Z} \quad \int_t^u \frac{1}{(x^2+bx+c)^i}\, dx =$$

0	$t = u$
unknown	$(t \neq u) \wedge$ $((i < 1) \vee (b^2 \geq 4c))$
$\dfrac{2}{\sqrt{4c-b^2}}\left(\tan^{-1}\left(\dfrac{(2u+b)}{\sqrt{4c-b^2}}\right) -\right.$ $\left.\tan^{-1}\left(\dfrac{(2t+b)}{\sqrt{4c-b^2}}\right)\right)$	$(t \neq u) \wedge$ $(i = 1) \wedge (b^2 < 4c)$
$\dfrac{(b+2u)}{(i-1)(4c-b^2)(u^2+bu+c)} -$ $\dfrac{(b+2t)}{(i-1)(4c-b^2)(t^2+bt+c)} +$ $\dfrac{(4i-6)}{(i-1)(4c-b^2)}\displaystyle\int_t^u \frac{1}{(x^2+bx+c)^{i-1}}\, dx$	$(t \neq u) \wedge$ $(i > 1) \wedge (b^2 < 4c)$

Figure 3. Entries for "Quadratic" Queries.

How to Find Symmetries Hidden in Combinatorial Problems

Noriko H. Arai   Ryuji Masukawa

Abstract. *In this paper, we describe a propositional system, "Simple Combinatorial Reasoning" as a ground theorem prover. We adopt tableau and DLL expressed as sequent calculi for the base systems and implement a symmetry rule on it. We show that our prover successfully finds symmetries in many elementary combinatorial problems, which are known to be exponentially hard for resolution and tableau, and automatically produces polynomial-size proofs. Furthermore, our prover distinguishes those formulas which contain symmetries and those which do not with high possibility without losing much time. As a result, the performance of our prover on randomly generated formulas is as good as that of existing resolution or tableau provers.*

1 Introduction

Since Haken found the first hard example for resolution [13], many others were added to the list of tautologies which require superpolynomially long proofs for resolution and analytic tableau [7]. Actually most of the interesting combinatorial problems were found hard for these proof systems. It was depressing news for the society of automated theorem proving since many automated provers adopt either resolution or analytic tableau as their engines. However, it was a quite natural consequence when we ponder how we human being reason. We use different reasoning for different types of problems; algebraic approach to the problems related to counting or linear algebra, combinatorial approach to those related to graphs. If we always take only one approach, which is purely logical analysis when we adopt resolution and analytic tableau, it is very likely that we end up with exponentially long proofs.

What we suggest in this paper is to give-up "only-one" approach and to adopt different approaches to different types of problems in ground theorem proving. The prover we designed in this paper features two theorem provers. One is DLL expressed as a sequent calculus and the other is Simple Combinatorial Reasoning. Introduced by Arai (1996), Simple Combinatorial Reasoning is a propositional proof system designed exclusively for combinatorial problems. It features the symmetry rule which allows the exploitation of symmetries present in a problem. It polynomially proves the pigeonhole principle, the mod-k principles, Bondy's theorem, Clique-Coloring problems and many other combinatorial problems; all of them are known to be hard for both resolution and tableau.

Although quite a number of researchers share Slaney's opinion: "I consider symmetry to be one of the most important topics of current research in ground theorem proving" [15], not much effort was done to design a theorem prover exploiting symmetries.[1] One reason why people were not so enthusiastic in adopting symmetries in the real prover is that finding symmetries seemed to be as time consuming as exhaustive search anyway. When a formula contains n variables, the most naive program to search for symmetries will check all the permutations on n variables; $n!$ permutations all together. The second reason is that symmetry rules does not seem to make any progress in shortening proofs for randomly generated formulas; the implementation of symmetry rules do not seem to improve the average time complexity. To make the situation worse, it was proved that finding a permutation of the longest orbit in a given formula is NP-complete, and asking two given formulas are symmetric is as hard as the graph isomorphism problem, which is conjectured not in the class P [11]. However, we should not misuse this evidence to conclude that symmetry rules are effective only in theory, but not in practice. It only tell us that we cannot always find the symmetries hidden in formulas, and symmetries will not give us much when we focus on randomly generated formulas.

In this paper, we set our goal to design a ground theorem prover so that

1. it finds symmetries in a propositional formula as long as a human being can find the symmetries in the corresponding first order formula, and

2. it can quickly decide whether a symmetry rule is worth trying; it distinguishes formulas with a lot of symmetries from those without them.

---

[1] The authors were notified that there exists a prover exploiting symmetries [9]. Unfortunately, the article is not obtainable for the authors at the present time. Hence, we are not able to compare how our system is different from theirs in this article.

Notice that our goal does not contradict any of the pessimistic evidences.

The symmetry rules can be added to resolution, tableau or sequent calculus. Since Krishnamurthy first pointed out that the symmetry rules is effective to shorten resolution refutations, researchers have focused on the symmetry rules in resolution [5, 6, 14]. It was Benhamou and Sais who first presented an algorithm to implement the symmetry rules in resolution [5]. Their strategy was to find a permutation of the largest orbit in the given formula before the machine started the resolution procedure. It was pointed out in [11] that finding a permutation of the longest orbit is an NP-complete problem, but Benhamou and Sais allowed the machine to backtrack only for fixed amount of time, therefore their algorithm has polynomial-time complexity. They demonstrated that their SLDI resolution prover with symmetry rules can automatically produce polynomial-size refutations for the pigeonhole principle. Unfortunately, the Benhamou-Sais algorithm did not overwhelm other techniques without the symmetry rules mainly for the following two reasons.

1. The B-S algorithm heavily depends on the form of the input clauses, and it does not work when we disturb its symmetries by throwing in some unnecessary clauses or additional variables.

2. It does not feature any precandition whether we should run the subroutine to find symmetries; it always tries to find symmetries.[2] Consequently, we end up with poor average time-complexity although it may run very fast for a small class of interesting formulas.

To overcome these deficiencies, we implemented our prover as a sequent-calculus-type backward search prover, called Godzilla. Godzilla does not try to find symmetries in the input formula, but it finds them while breaking down the formula. Godzilla almost always finds symmetries and produces proofs of size linear to the size of inputs for the pigeonhole principle, the mod-k principle, the clique-coloring problem and other test formulas without increasing the time-complexity much.

However, the performance of the original Godzilla turned out to be much poorer than existing DLL provers for randomly generated formulas. One reason is that DLL is theoretically faster than tableau, and another is that Godzilla did not use any heuristic favor for randomly generated 3-CNF formulas. Another criticism of Godzilla was that the performance of Godzilla on combinatorial formulas seemed to rely on how nicely the input formulas were formulated. In this paper, we adopt both tableau and DLL as the basis for the new Godzilla so that we can choose either

---

[2]A hard example for the B-S algorithm can be found in [4].

of them according to the conditions satisfied by the input formula. As a good side-effect, the new Godzilla proves some combinatorial problems which the old model was not able to produce short proofs. We discuss the detail in Section 4. As a result, the performance of Godzilla is improved considerably. We also experiment to determine whether or not Godzilla can appropriately find symmetries when we shuffle the input clauses.

This paper is organized as follows. In Section 2, we analyze proofs for elementary combinatorial problems. In Section 3, we define a deterministic algorithm to simulate elementary combinatorial proofs line by line, and implement it as a theorem prover, Godzilla. In Subsection 4.1, we demonstrate how Godzilla produces proofs for the set of clauses of size n on n variables, the pigeonhole principle and the clique-coloring problem, which surprisingly resemble to human proofs. In Subsection 4.2, we shuffle the input clauses of the pigeonhole principle and see whether Godzilla can still find symmetries. In Subsection 4.3, we examine the performance of Godzilla on randomly generated formulas.

2 Simple Combinatorial Proofs

In this section, we informally define what *elementary combinatorial proofs* are, and discuss how to find the symmetries hidden in problems and how to exploit them to obtain short proofs. By analyzing proofs for simple combinatorial problems step by step, we try to extract the reason why these problems are so straightforward for us while they are exponentially hard for many automatic provers.

The pigeonhole principle is one of the most elementary combinatorial principles. The pigeonhole principle states that there is no 1-1 mapping from the set of $n + 1$ objects into the set of n objects. This principle is known to be hard for tableau, resolution and even for bounded depth Frege systems, although the truth of the principle is clear for us. The best thing we can do to prove the principle in resolution is to go over all the possible cases, $n!$ cases all together, that is slightly better than the truth table.

An elementary proof of the pigeonhole principle uses mathematical induction on the number, n, of objects in the domain; we assume that the pigeonhole principle holds for n, and show that it also holds for $n + 1$.

(Informal proof of the pigeonhole principle)

Let f be a mapping from $\{0, \ldots, n + 2\}$ to $\{0, \ldots, n + 1\}$. Without loss of generality, we can assume that $f(n + 2) = n + 1$. If there exists an $i \neq n + 2$ such that $f(i) = n + 1$, we are done. Suppose otherwise. Then the function f restricted to $\{0, \ldots, n + 1\}$ is a mapping to $\{0, \ldots, n\}$. By the induction hypothesis, it is not one-to-one, and so is not f (q.e.d.).

The novelty of the proof given above is the line, "Without loss of generality ...". Here, we understand that the situation of $f(n + 2) = i$ $(i = 0, \ldots, n)$ is merely a variant of the situation of $f(n + 2) = n + 1$; we save time by representing (exponentially) many cases by just one case.

We give another example which has a slightly different proof structure. We define $\Pi(n)$ as the set of all clauses of length n in n variables. $\Pi(n)$ is an unsatisfiable set of clauses. D'Agostino proved that this problem is hard for analytic tableau [10]: it requires a proof of size at least $n!$, which is superpolynomial of 2^n. The unsatisfiability of $\Pi(n)$ is informally proved as follows.

(Informal proof of unsatisfiability of $\Pi(n)$)
Let $p_1, \ldots, p_n$ denote the list of variables appearing in $\Pi(n)$. $\Pi(n)$ is unsatisfiable if and only if both $\Pi(n)|_{p_1=T}$ and $\Pi(n)|_{p_1=\perp}$ are unsatisfiable. However, both of the formulas are equivalent to $\Pi(n-1)$. By the induction hypothesis, $\Pi(n-1)$ is unsatisfiable. (q.e.d.)

In this proof, again, we understand that $\Pi(n)$ with the assumption p_1 being unsatisfiable and that with the assumption p_1 being false are isomorphic in structure. Consequently, we represent exponentially many cases by just one case.

The main structure of these proofs is summarized as follows.

1. The statement to be proved is a big disjunction of subcases,

$$\bigvee_{1 \leq i \leq h} A_i$$

   where A_i and A_j $(1 \leq i, j \leq h)$ are isomorphic each other.

2. The formula A_1 is reducible (using pure logic) to an induction hypothesis, or A_1 has a short proof.

We define *elementary combinatorial proofs* to be those having a structure satisfying conditions (1) and (2) given above. Many combinatorial principles are known to have elementary combinatorial proofs; the mod-k principle, the non-unique endnode principle and Bondy's theorem are a few examples.

Now we try to simulate elementary combinatorial proofs in the propositional setting. We assume that formulas are expressed as CNF; the input formula is expressed as a set of clauses; $A = C_1 \wedge \cdots \wedge C_n$ and

$C_i = l_1^i \vee \cdots \vee l_{m_i}^i$. The first task is to understand the given formula, A, as a big disjunction of subcases.

In the case of the pigeonhole principle, there exists a clause C_i $(1 \leq i \leq n)$ such that

$$
\begin{aligned}
A \quad &= \quad (C_1 \wedge \cdots C_{i-1} \wedge l_1^i \wedge C_{i+1} \wedge \cdots \wedge C_n) \\
&\vee \cdots \vee \quad (C_1 \wedge \cdots C_{i-1} \wedge l_{m_i}^i \wedge C_{i+1} \wedge \cdots \wedge C_n)
\end{aligned}
$$

and each $(C_1 \wedge \cdots C_{i-1} \wedge l_j^i \wedge C_{i+1} \wedge \cdots \wedge C_n)$ is isomorphic to the induction hypothesis.

The proof structure for $\Pi(n)$ is different; there exists a variable p such that A with the assumption p and that with the assumption $\bar{p}$ are both isomorphic to the induction hypothesis.

The first kind of reasoning is most naturally expressed as tableau-like sequent calculus, on the other hand DLL-like sequent calculus is more suitable to express the second kind.

The old prover we designed to simulate elementary combinatorial proofs was equipped only with tableau. Hence, it failed to find the second kind of symmetries discussed above. To overcome this deficiency, we design our prover so that it can choose either tableau or DLL according to the type of the input formula.

3   Theorem Prover: Godzilla

3.1   Algorithm

In this subsection, we implement a ground theorem prover, *Godzilla*, to simulate elementary combinatorial proofs discussed in the previous section. The algorithm of Godzilla consists of three parts. The first part is a tableau-like sequent calculus, the second is a DLL-like sequent calculus, and the third part takes care of the restricted permutation rules.

Each of the first and the second part consists of two subparts: *Simplification* and *Branching*. Simplification consists of three subroutines. The first subroutine checks whether or not a given set of clauses contains an axiom. We delete unnecessary clauses as much as possible in the second subroutine. The third subroutine is the unit propagation (unit resolution). During unit propagation, the set of clauses is reordered. Branching divides the given set of clauses into several subsets. Obviously, Branching is the main cause to blow-up the size of proofs.

The third part of the algorithm checks whether or not two given formulas are isomorphic. This procedure is called *Musical-Chairs*. It is quite important not to play Musical-Chairs when it is hopeless that two given formulas are isomorphic, otherwise the average performance of our prover will be quite poor compared to the existing Davis-Putnam based theorem prover. For this purpose, we inserted a procedure called *Checker* to examine whether we should try Musical-Chairs or not.

Now we explain the flow of the algorithm. For a technical reason, we describe the algorithm so that the machine produces proofs for several sets of clauses stored in a database. Each set of clauses is expressed as a sequence of clauses, called a *sequent*. Each sequent S is labeled with the number of clauses in S, denoted by $len(S)$, and a sequence of integers $seq(S)$ of length $len(S)$ such that the ith element in $seq(S)$ is the size of the ith clause in S. We call $seq(S)$ the characteristic sequence of S. Suppose that a sequent S is of the form

$$p_1 p_2 p_3, p_4 p_5, \bar{p}_1 \bar{p}_4, \bar{p}_3 \bar{p}_5.$$

Then, $len(S) = 4$ and $seq(S)$ is $3, 2, 2, 2$.

We first run the subroutine *Simplification* for all the sequents in the database in pararell. Next, we send the database to *Musical-Chairs*. In *Musical-Chairs* new databases are formed. Finally, we send the new databases to *Branching*, and back to *Simplification*.

1. First, check if all the elements in the database are proper sets of clauses.

2. (*Simplification*) Simplification consists of three subroutines.

   (a) (*Subroutine 1*) For each sequent S in the database, check whether or not S contains an axiom: for some literal l, both $\{l\}$ and $\{\bar{l}\}$. If S contains an axiom, write "S is unsatisfiable." If S is an empty sequent, stop the whole procedure immediately and output "(The input is) satisfiable." If not, go to next subroutine.

   (b) (*Subroutine 2*) For each S in the database, find a literal l such that there exists a clause containing l but there is none containing $\bar{l}$. Delete all the clauses containing l. If there is no such l, go to the next subroutine.

   (c) (*Unit Propagation*) For each sequent S in the database, find a unit clause $\{l\}$. Delete the clauses containing l. Move all the clauses containing $\bar{l}$ to the head of S, delete the occurrences of $\bar{l}$, and go back to Subroutine 1. Otherwise, this is the end of Simplification, and go to *Checker*.

3. In *Checker*, we divide the given database into disjoint subdatabases.

   (a) (*Checker 1*) First, partition the given database into databases consisting of the same length of sequents. When a database consists of a single sequent, send it to *Branching*.

   (b) (*Checker 2*) Next, partition the given database into databases consisting of sequents so that S and S' are in the same database iff $seq(S) = seq(S')$. When a database consists of a single sequent, send it to *Branching*.

4. (*Musical Chair*) Now we are looking at a database consisting of sequents having the same characteristic sequences. Pick two sequents S_1 and S_2 in the database. (Note that the number of combinations is $O(n^2)$.) S_1 and S_2 are expressed as follows.

$$C_1, \ldots, C_n \ (= S_1)$$
$$D_1, \ldots, D_n \ (= S_2)$$

$size(C_k) = size(D_k)$ for every $1 \le k \le n$. Suppose that C_1 is a clause of the form $l_1 \cdots l_m$ and D_1 is of the form $t_1 \cdots t_m$. Without loss of generality, we can assume that C_1 and D_1 are disjoint sets of literals. Define a permutation π by a product of transpositions as follows.

$$\pi = \left(\begin{array}{ccc} l_1 & t_1 \end{array} \right) \cdots \left(\begin{array}{ccc} l_m & t_m \end{array} \right)$$

Extend π so that $\pi(l) = t$ if $\pi(\bar{l}) = \bar{t}$. Rename literals in S_1 according to π. If $\{\pi(C_2), \ldots, \pi(C_n)\} = \{D_2, \ldots, D_n\}$ as sets of clauses, then delete $C_1, \ldots, C_n$ from the database because it is reducible to $D_2, \ldots, D_n$ by using a symmetry rules. Otherwise, move all the clauses containing a literal in $(C_1 - \pi(C_1))$ or its negation to the end of the sequent in S_1. On the other hand, move the clauses which contain a literal in $(\pi(D_1) - D_1)$ or its negation to the end of the sequent in S_2. Send the two obtained sequents back to *Musical-Chairs*. If we are still playing on the same two sequents after running this procedure n times, pick a different combination of sequents. When we finish checking all the combinations, it is the end of *Musical-Chairs*. For each sequent left in the database, form a new database consisting of the sequent, and send them to *Branching*.

5. When we receive a database from *Musical-Chairs*, it always consists of a single sequent S. *Branching* has two subroutines called *Tableau* and *DLL*. In this paper, we adopt the following condition as our heuristic

to decide whether we should apply *Tableau* or *DLL*. If there exists a clause C such that no literal in C appears in the other clauses in S, go to *Tableau*. Otherwise go to *DLL*.

(a) (*Tableau*) Without loss of generality, we assume that S is of the form

$$C_1, C_2, \ldots, C_n$$

where no literals in C_1 appears in other clauses. Delete S from the database, and add new sequents

$$\{l\}, C_2, \ldots, C_n$$

for each $l \in C_1$. Send the new database back to *Simplification*.

(b) (*DLL*) Pick a variable p of most occurrences in S. Delete S from the database, and add two new sequents $\{p\} \cup S$ and $\{\bar{p}\} \cup S$, to the database. Send the new database back to *Simplification*.

6. If every sequent in every database is unsatisfiable, output "(The input is) unsatisfiable."

When we input a set of unsatisfiable clauses, $\{C_1, \ldots, C_n\}$, Godzilla produces a elementary combinatorial proof expressed as a directed acyclic graph so that every leaf of P is labeled by an axiom. When we input a set of satisfiable clauses, $\{C_1, \ldots, C_n\}$, Godzilla stops immediately when it finds a satisfying valuation.

Time-Complexity All of the subroutines are accomplished in time $O(n^4)$ where n is the size of the input. Hence, if the size of the obtained proof is polynomially bounded, the time complexity to obtain the proof is also polynomially bounded. Hence, we can assess the efficiency of Godzilla by the size of proofs generated by Godzilla. The obvious upper bound for the size of proof is $k \cdot 2^n$, where n is the number of variables contained in a given formula, and k the size of the formula.

Memory The number of sets of clauses stored in the memory is bounded by (max clause length)×(number of variables). The size of each set of clauses is bounded by that of the input set.

4 Experimental Results

4.1 How Godzilla Simulates Human Reasoning

We first demonstrate how Godzilla acts on combinatorial problems; $\Pi(n)$, the pigeonhole principle, and the clique-coloring problem.

The *clique-coloring problem*, denoted by k-Test(n), states that if a graph contains a k-clique, the graph cannot be properly colored with $(k-1)$ different colors. To express the clique-coloring problem in the propositional calculus, we introduce three types of variables; one to express the clique function, one to express whether there exists an edge between given two vertices, and another to express the coloring function. In this model, how we should permute "coloring variables" is determined by the permutation of "edge variables," which is determined by that of "clique variables"; finding an appropriate permutation for the clique-coloring problem is a lot harder than that for the pigeonhole principle.

Figures 1, 2, and 3 show comparisons of the performance of Godzilla with and without permutation rules for $\Pi(n)$, PHP(n) and $(n-1)$-Test(n) in CPU time.[3]

As mentioned in the previous section, the cost of Musical-Chairs has time complexity $O(n^4)$.

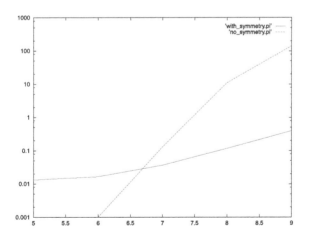

Figure 1. Godzilla w/ vs. w/o symmetries on $\Pi(n)$.

---

[3]Graphs show in logscale. Run times are in seconds and are for C version of Godzilla running on a AMD-K6-2 300MHz processor with 128 MB ram.

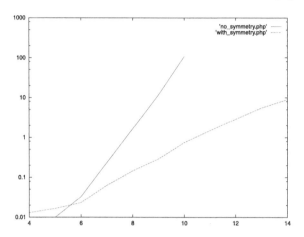

Figure 2. Godzilla w/ vs. w/o symmetries on PHP (n).

Godzilla almost always finds necessary permutations for these elementary combinatorial problems. Furthermore, Godzilla applied DLL for $\Pi(n)$ and Tableau for PHP(n) and $(n-1)$-Test(n), which were appropriate decisions. The number of nodes in proofs for $\Pi(n)$ produced by Godzilla is about square root of that produced by its old model.

Table 1 shows the number of leaves of the proofs generated by Godzilla for these problems.

4.2 Can Godzilla Find Symmetries when Input Clauses are Shuffled?

One of the main criticism of the old Godzilla was that Godzilla seemed to find symmetries only when the input was formulated nicely. Table 2 shows how the size of proofs increases when we shuffle the order of clauses in the pigeonhole principle. We varied the number of pigeons from 5 to 13. For each n, we ran Godzilla on 20 shuffled PHP(n) and took the average

	$n=3$	$n=4$	$n=5$	$n=6$	$n=7$	$n=8$	$n=9$
$\Pi(n)$	2	2	2	2	2	2	2
PHP(n)	2	2	2	2	2	2	2
Test(n)	2	2	4	2	4	4	4

Table 1. Number of leaves in proofs; Godzilla.

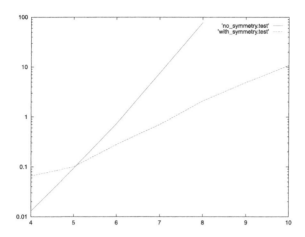

Figure 3. Godzilla w/ vs. w/o symmetries on $(n-1)$-Test(n).

number of leaves in the proofs generated by Godzilla. The result shows that
the chance for Godzilla to find symmetries is much worse than the results
in the previous subsection. It should be worth noting that even if a prover
fails to recognize two formulas are isomorphic only 1 out of 100 cases, the
size of proofs may still blow-up exponentially. We need more techniques
to improve the ability to find symmetries when the input formula is not
formulated nicely. Analyzing the proof produced by Godzilla on a shuffled
pigeonhole principle with 7 pigeons, we observed that when the sequents
become longer, it is hard for Godzilla to recognize two given sequents are
isomorphic after shuffling; it is hard to find symmetries in the beginning of
the proof. However, the reordering process in the unit propagation helped
Godzilla to find the symmetries, and the possibility to find symmetries
increases towards the end of the proof.

When we extend Godzilla so that it can simulate course-of-values induction, we obtain better performance in finding symmetries for shuffled for-

	$n = 4$	$n = 5$	$n = 6$	$n = 7$	$n = 8$	$n = 9$
DLL	6	24	120	720	5040	40320
Godzilla	2	4	14	62	353	2278
course-of-values ind.	2	3	4	4	7	9

Table 2. Number of leaves for shuffled PHP (n); Godzilla with vs. without
symmetry rules.

mulas or more complicated problems. However, in return, the performance for randomly generated 3-CNF drops severely because of the search-space blow-up.

4.3   How Godzilla Acts on Randomly Generated Formulas

It is a key for the success of Godzilla not to increase time-complexity when it is attacking a tautology having no combinatorial model. When the given formula does not have any combinatorial model, there is little hope that the formula contains many symmetries. Randomly generated formulas are typical examples. As described in Section 3, Godzilla is endowed with a subroutine called *Checker* so that it stops immediately when the given sequent seems to be asymmetric. Our preliminary experiments showed that when randomly generated 3-CNF formulas are broken into several sequents by Branching, only 3 cases out of 1000 passed Checker; Checker seems to be quite effective to detect which sequents contain symmetries and which do not.

Figure 4 shows a comparison of Godzilla with and without symmetry rules on 50 variable randomly generated 3-CNF in CPU time. We varied the number of clauses from 160 to 300. Godzilla only lost 16% of time by having the symmetry rules.

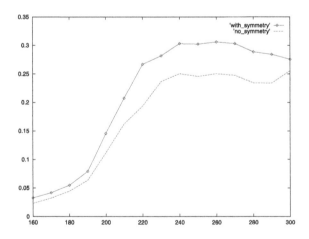

Figure 4. Godzilla w/ vs. w/o symmetries on random 3-CNF.

5 Conclusion

Our theoretical results show that permutation rules (or symmetry rules) have a dramatic impact on reducing the lengths of proofs for many combinatorial problems, which are hard for both resolution and tableau. Moreover, our experimental results show that finding symmetries in a given formula is not as hard as it was believed when we adopt the sequent calculus for the base system.

It is a key for the intellectual theorem proving how accurately and how easily the machine can recognize which field of mathematics a given problem falls in. Then, we can apply algebraic technique, for example the cutting planes, for algebraic problems, symmetry rules for elementary combinatorial problems, and common resolution for randomly generated 3-CNF's. In this paper, we used naive heuristics to distinguish whether we should apply the symmetry rules or not, that worked quite successfully in the restricted setting. We will need more delicate heuristic functions when we extend our technique to prove problems which have various types of mathematical models.

References

[1] N.H. Arai. Tractability of cut-free Gentzen type propositional calculus with permutation inference, *Theoretical Computer Science*, Vol. 170 (1996) 129-144.

[2] N.H. Arai. Tractability of cut-free Gentzen type propositional calculus with permutation inference II, to appear in *Theoretical Computer Science*.

[3] N.H. Arai. No feasible monotone interpolation for simple combinatorial reasoning, to appear in *Theoretical Computer Science*.

[4] N.H. Arai and A. Urquhart. Local symmetries in propositional logic. *Proc. of TABLEAUX 2000*, vol. 1947 of LNAI (2000) 40-51.

[5] B. Benhamou and L. Sais. Tractability through symmetries in propositional calculus. *Journal of Automated Reasoning*, Vol. 12 (1994) 89-102.

[6] W. Bibel. Short proofs of the pigeonhole formulas based on the connection method. *Journal of Automated Reasoning*, Vol. 6 (1990) 287-297.

[7] V. Chvátal and E. Szemerédi. Many Hard Examples for Resolution. *Journal of the Association for Computing Machinery*, Vol. 35 (1988) 759-768.

[8] J. Crawford, L. Auton. Experimental results on the crossover point in satisfiability problems. *Proc. of the 11th AAAI* (1993) 21-27.

[9] J. Crawford et al. Symmetry – breaking predicates for search problems. *Proc. of 5th International Conference on Principles of Knowledge Representation and Reasoning* (1996) 148-159.

[10] M. D'Agostino. Are tableaux an improvement on truth-tables?. *Journal of Logic, Language and Information*, Vol.1 (1992) 235-252.

[11] T. De la Tour and S. Demri. On the complexity or extending ground resolution with symmetry rules. *Proc. of the 14th IJCAI* (1995) 289-295.

[12] J. H. Gallier. *Logic for Computer Science*, John Wiley & Sons, New York (1987).

[13] A. Haken. The intractability of resolution. *Theoretical Computer Science*, 39 (1985) 297-308.

[14] B. Krishnamurthy. Short proofs for tricky formulas. *Acta Informatica*, Vol. 22 (1985) 253-275.

[15] J. Slaney. The crisis in finite mathematics: automated reasoning as cause and cure. In *CADE-12, Nancy*, ed. A Bundy, Springer Verlag, LNAI 814 (1994) 1-13.

Communication Protocols for Mathematical Services based on KQML and OMRS

Alessandro Armando   Michael Kohlhase
Silvio Ranise

Abstract. *In this paper we describe the first ideas for formalizing a communication protocol for mathematical services based on* KQML *(Knowledge Query and Manipulation Language) and* OMRS *(Open Mechanized Reasoning Systems). The claim is that the interaction level of a communication protocol for mathematical services can be relatively generic (hence* KQML *suffices), as long as the ontology of the computational behavior and internal state of the mathematical services is sufficiently expressive and concise (which we have in* OMRS*).*

The material presented in this paper is a first exploratory step towards the definition of the interaction level in OMRS*, supplies a concrete syntax based on the* OPENMATH *standard, and gives a semantics to communication of mathematical services in distributed theorem proving and symbolic computation environments.*

1   Introduction

It is plausible to expect that the way we do (conceive, develop, communicate about, and publish) mathematics will change considerably in the next ten years. The Internet plays an ever-increasing role in our everyday life, and most of the mathematical activities will be supported by mathematical software systems (we will call them *mathematical services*) connected by a commonly accepted distribution architecture, which we will call the *mathematical software bus*. We have argued for the need of such an architecture in [13], and we have in the meantime gained experiences with prototype systems (the MATHWEB software bus [14] (MATHWEB-

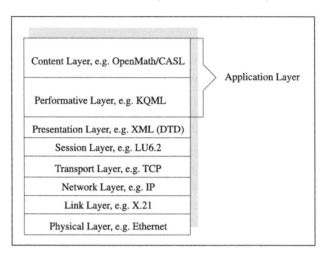

Figure 1. Artificial Communication: KQML and the OSI Reference Model.

SB) and the LOGIC BROKER ARCHITECTURE [5] systems); other groups have conducted similar experiments [15, 10] based on other implementation technologies, but with the same vision of creating a world wide web of co-operating mathematical services. In order to avoid fragmentation, double inventions and to foster ease of access it is necessary to define interface standards for MATHWEB.[1] In [13], we have already proposed a protocol based on the agent communication language KQML [12] and the emerging Internet standard OPENMATH [8, 28] as a content language (see Figure 1). This layered architecture which refines the unspecific "application layer" of the OSI protocol stack is inspired by the results from agent-oriented programming [19, 18], and is based on the intuition, that all agents (not only mathematical services) should understand the agent communication language, even if they do not understand the content language, which is used to transport the actual mathematical content. The agent communication language is used to establish agent identity, reference and—in general—model the communication protocols. The content language is used for transporting the mathematical content.

In this paper, we refine the communication protocol proposed for MATH-WEB based on the experiences gained since [13]. The problem with that

---

[1] We will for the purposes of this paper subsume all of the implementations by the term MATHWEB, since the communication protocols presented in this paper will make the constructions of bridges between the particular implementation simple, so that the combined systems appear to the outside as one homogenous web. There is a joint effort underway to do just that at http://www.mathweb.org

proposal was that pure OPENMATH [8] is too weak as a content language, since it is geared towards the representation of mathematical objects, which is sufficient for content communication among symbolic/numeric computation services, but not for reasoning services. In [22] the second author has presented an extension ℳDOC (OPENMATH Documents) of OPENMATH by primitives for document structure, theory management, and proofs, arriving at a (structured) specification language for symbols, definitions, theorems, theories, etc. which make up the content reasoning services need to communicate about. Yet, this is still not sufficient to specify the *interaction* of mathematical services on MATHWEB; this is where the KQML kicks in. In fact, KQML was developed exactly for the purpose of agent interaction, assuming that the content layer has already been specified.

However, in order to build a protocol based on something as general as the KQML we need a way to specify the state and the strategies of the mathematical services themselves. This is exactly what the OMRS framework has been developed for: to supply a specification framework for mathematical software systems. In OMRS, a mathematical software system is structured in three layers: the *logic* layer (specifying the deductive machinery), the *control* layer (specifying the strategies for controlling inference), and the *interaction* layer (specifying the interaction capabilities of the mathematical software system with others and with humans).[2] The **main claim** of this paper is that the interaction level of a communication protocol for mathematical services can be relatively generic (hence KQML suffices), as long as the ontology of the computational behavior and internal state of the mathematical services is sufficiently expressive and concise (which we have in OMRS).

The plan of the paper is as follows. In Section 2, we briefly review KQML and present a simple ontology for reasoning services in MATHWEB (Subsection 2.1). In Section 3, we hint the main concepts of the OMRS framework. Then, in Section 4, we put together KQML and OMRS to validate our claim on a significant example. In Section 5 we discuss a migration path for the existing MATHWEB software bus implementations towards a joint architecture based on the protocol presented in this paper. Finally, in Section 6, we draw some conclusions.

---

[2]The OMRS framework—initially conceived to specify deduction systems—has been extended to support the specification of computer algebra systems in [6]. For the sake of simplicity in this paper we use the original OMRS framework.

2   The KQML and MathWeb

KQML, the Knowledge Query and Manipulation Language,[3] is a language and protocol for exchanging information and knowledge among software agents. It is both a message format and a message-handling protocol and can be used as a language for an application program to interact with an intelligent system or for two or more intelligent systems to share knowledge in support of cooperative problem solving.

KQML focuses on an extensible set of performatives, which defines the permissible operations that agents may attempt on each other's knowledge and goal stores. The performatives comprise a substrate on which to develop higher-level models of inter-agent interaction such as contract nets and negotiation. Following [24], KQML performatives can be modeled as actions which modify the cognitive states of the agents. The cognitive states of the agents can be modeled by means of the predicates:

- $bel(A, X)$ asserts that X is true for agent A (where X is a statement about the application domain), or equivalently that X is in the (virtual) knowledge base of A.

- $know(A, Y)$ asserts that Y is known to be true by A. (Here and below Y is a statement about the cognitive states of the agents).

- $want(A, Y)$ asserts that A desires the state described by Y to occur.

- $intend(A, Y)$ asserts that A has every intention of achieving the state described by Y.

Notice that while the meaning of *know*, *want*, and *intend* is fixed, the meaning of *bel* depends on the application. The semantics of KQML performatives is given in terms of the preconditions and postconditions describing the applicability conditions and the effect of the performatives respectively. A (simplified) account of the semantics of the `ask-if` and `tell` performatives is given in Figure 2. (The specification of the `deny` performative can be obtained from that of `tell` by replacing every occurrence of $bel(A, X)$ with $\neg bel(A, X)$.)

In addition, KQML provides a basic architecture for knowledge sharing through a special class of agents called communication facilitators which coordinate the interactions of other agents.

---

[3]More information can be obtained from `http://www.cs.umbc.edu/kqml`.

ask-if(A,B,X)
Pre: $want(A, know(A, \mathcal{Y}))$
Post: $intend(A, know(A, \mathcal{Y}))$
$know(B, want(A, know(A, \mathcal{Y})))$

tell(A,B,X)
Pre: $bel(A, X)$
$know(A, want(B, know(B, \mathcal{Y})))$
$intend(B, know(B, \mathcal{Y}))$
Post: $know(A, know(A, bel(A, X)))$
$know(B, bel(A, X))$

Legend: $\mathcal{Y}$ stands for $bel(B, X)$, $bel(B, \neg X)$, or $\neg bel(B, X)$.

Figure 2. Semantics of KQML performatives.

2.1 Reasoning Services in MATHWEB

A MATHWEB reasoning service is a mechanized reasoning system that tries to determine whether a given logical formula is valid (satisfied by all models), satisfiable (satisfied by some models), or unsatisfiable (not satisfied by any models). Automated theorem provers typically try to determine validity (theorem-hood) of a formula F^4 by finding a proof for F, or trying to refute the satisfiability of its negation $\neg F$. Model generators try to follow a dual approach, they try to construct a model for a formula F; if they succeed, then F is shown to be satisfiable, if they fail, exhausting all possibilities, then F is unsatisfiable.

In MATHWEB we use the simple general hierarchy of states of mechanized reasoning systems to interact with the systems in KQML, which is shown in Figure 3. With this ontology, it is simple to communicate with all sorts of automated reasoning systems in KQML. For instance, the messages in Figure 4 are the normal way to request a judgment about the theorem-hood of a formula F.

Upon receiving the first message, the service atp at mathweb.org,[5] will try to determine the semantic status of the formula F by, e.g., concurrently starting one or several theorem provers and model generators in order to achieve the determined status. Upon receiving the subscribe message, it responds with an appropriate message (as if processing the ask-if message, most likely with a tell message) immediately after the status has been determined.

Querying an automated theorem proving system for a proof is similar, only employing an ask-one message using the more concrete statuses model, proof, counter-model, counter-proof, or mcm. These are defined in a special OPENMATH content dictionary reasys (CD, see the refer-

---

[4]This will in general be of the form $A_1 \wedge \ldots \wedge A_n \Rightarrow C$, where the A_i are the assumptions (e.g. supplied by some background theory) and C the conclusion.

[5]It speaks the XML representation of KQML described in this paper, as we see by the prefix xml-kqml, so the representation of the messages is appropriate.

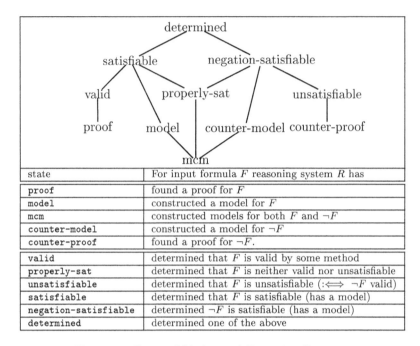

state	For input formula F reasoning system R has
proof	found a proof for F
model	constructed a model for F
mcm	constructed models for both F and $\neg F$
counter-model	constructed a model for $\neg F$
counter-proof	found a proof for $\neg F$.
valid	determined that F is valid by some method
properly-sat	determined that F is neither valid nor unsatisfiable
unsatisfiable	determined that F is unsatisfiable ($:\Longleftrightarrow \neg F$ valid)
satisfiable	determined that F is satisfiable (has a model)
negation-satisfiable	determined $\neg F$ is satisfiable (has a model)
determined	determined one of the above

Figure 3. States of Mechanized Reasoning Systems.

ences in the OPENMATH symbols OMS in Figure 4), which is available from http://www.mathweb.org/omdoc/cd/reasys.ocd.

3   An Overview of OMRS

An OMRS specification consists of three layers: the *logic* layer (specifying the assertions manipulated by the system and the elementary deductions upon them), the *control* layer (specifying the inference strategies), and the *interaction* layer (specifying the interaction of the system with the environment). Notice that this layering allows for an additional and complementary way to structure the specifications w.r.t. the standard approach based on modularity. This domain-specific feature of the OMRS specification framework is fundamental to cope with the complexity of functionalities provided by state-of-the-art implementations.

The Logic Layer.   The logic layer of an OMRS specification describes the assertions manipulated by the system and the elementary deduction

```
<achieve sender="A"
        receiver="k qml-xml://mathweb.or g#atp"
        reply-with="id1"
        language="OpenMath">
  <OMOBJ><OMA><OMS cd="reas ys" name="determined"> F</OMA></OMOBJ>
</achieve>

<subscribe sender="A"
        receiver="k qml-xml://mathweb.or g#atp"
        reply-with="id2"
        language="OpenMath">
  <ask-if sender="A"
        receiver="k qml-xml://mathweb.or g#atp"
        reply-with="id3"
        language="OpenMath">
    <OMOBJ><OMA><OMS cd="reas ys" name="valid"> F</OMA></OMOBJ>
    <OMOBJ><OMA><OMS cd="reas ys" name="proper-sat"> F</OMA></OMOBJ>
    <OMOBJ><OMA><OMS cd="reas ys" name="unsatisfiable"> F</OMA></OMOBJ>
  </ask-if>
</achieve>
```

Figure 4. Querying a Reasoning System for Semantic Status of F.

steps the system performs upon such assertions. For example, a resolution-based theorem prover may manipulate first-order clauses by resolving and factorizing them. As another example, a linear arithmetic decider may manipulate polynomial inequalities by cross-multiplications and sums. At the logical level, the computations carried out by the system amount to constructing and manipulating structures consisting of assertions connected through elementary deduction steps (like proof trees). The key concept of the logic layer is that of reasoning theory. Roughly speaking, a *reasoning theory (RTh)* [16] consists of a set of *sequents* (i.e., assertions) and a set of *inference rules* over such sequents. An RTh defines a set of *reasoning structures*, i.e. graphs labeled by sequents and rules. The notion of reasoning structure generalizes the standard concept of derivation so as to capture, e.g., provisional reasoning and sub-proof sharing.

The Control Layer. Most real-world systems carry out their control strategies by making use of non-logical information, used exactly for control purposes. Examples of such control information are some history about how an assertion was produced, the number of times a certain inference step has been applied, the order in which some assertions must be selected for applying some reasoning steps, etc. Control information is used and modified during computation, at the same time as logical inferences are performed. The control layer of an OMRS specifies how reasoning systems manipulate the logic information, i.e., which strategies are used to select and apply the inference steps at each point of the computation. The concept of *Annotated Reasoning Theories* formalizes the control layer: It accounts for the simultaneous manipulation of logic and control information. More precisely, an annotated reasoning theory consists of a reasoning theory and an erasing mapping. The sequents of the reasoning theory associated to the annotated reasoning theory are (annotated) sequents over an extended syntax which encode both logic and control information; the inference rules specify how such information is manipulated by the reasoning system. Finally, the erasing mapping specifies what is the logical content of the annotated sequents. The interested reader is urged to see [1] for a complete discussion of the control layer of the OMRS framework.

Both reasoning theories and annotated reasoning theories can be glued together yielding composite reasoning theories and annotated reasoning theories respectively.

The Interaction Layer. The OMRS interaction layer specifies the interaction of the reasoning system with the environment. At present only exploratory work has been carried out on this (see, e.g., [29, 3]) and a rigorous development of the interaction layer is part of the future work. The work described in this paper is a first exploratory step towards the definition of the interaction level in OMRS.

4   Constraint Contextual Rewriting as a Case Study

We show the applicability of our ideas by specifying a distributed version of Constraint Contextual Rewriting (CCR, for short) [2, 4]. We do this by first providing an OMRS specification of CCR (adapted from [1]) and then by describing an agent-oriented architecture for CCR obtained by turning the decision procedure into an autonomous agent which interacts with the simplifier via KQML performatives. We will show that the OMRS specification plays a fundamental role in the specification of the agent-oriented architecture.

CCR extends traditional conditional rewriting by exploiting the functionalities of a decision procedure as described in [4, 2]. We illustrate CCR by means of a simple example. Let us consider the problem of simplifying the clause $a < 1 \vee f(a) = a$ using the following fact as a conditional rewrite rule:

$$x > 0 \Rightarrow f(x) = x \tag{10}$$

where $<$ and $>$ are the the standard 'less-than' and 'greater-than' relations over the integers (resp.) and f is an uninterpreted function symbol. The basic step of the simplification process is to select a literal (called the *focus literal*) from the clause and start rewriting it, while assuming the negation of the remaining literals (called the *context*). For the clause above, let $f(a) = a$ be the focus literal and $\{a \not< 1\}$ be the context. Application of (10) turns the focus literal into the identity $a = a$, under the proviso that the instantiated condition, namely $a > 0$, is entailed by the context. This can be established by means of a decision procedure for linear arithmetics by asking the decision procedure to check the satisfiability of the sets of literals obtained by adding the negation of the condition to the context, namely $\{a \not> 0, a \not< 1\}$. The simplification activity concludes that the identity $a = a$ is *true* by rewriting.

4.1   An OMRS Specification of CCR

An OMRS specification of CCR can be given by an annotated reasoning theory resulting from the combination of three annotated reasoning theories[6] each modeling a distinct reasoning module: the top level simplification loop (simp), the rewrite engine (cr), and the decision procedure (cs). The functional dependencies between the modules are depicted in Figure 5 (a). simp takes a clause (cl) and returns a simplified clause (cl'). cr performs conditional rewriting on the input literal (l) by using cs as rewriting context and returns a rewritten literal (l'). cs takes a conjunction of literals cnj and a context cs as input and returns a new context cs' obtained by extending cs with the literals in cnj.

The annotated reasoning theory for the overall simplification activity is the following (for details, see [1]). The sequents have the following forms:

- $cl \rightarrow_{\mathsf{simp}} cl'$ asserts that clause cl' (modeled as a finite set of literals) is the result of simplifying clause cl,

- $cs :: l \rightarrow_{\mathsf{cr}} l'$ asserts that literal l' is the result of rewriting l using cs (also called *constraint store*) as context,

- $cnj :: cs \rightarrow_{\mathsf{cs}} cs'$, asserts that cs' is the result of extending cs with the literals in cnj, cs-init(cs), asserts that cs is the "empty" constraint store, cs-unsat(cs) asserts that cs is an inconsistent (w.r.t. the theory decided by the decision procedure) constraint store, and

---

[6]For the lack of space we confine ourselves to specifying the control layer and provide an informal explanation of the logic content. The interested reader may consult [1] for the details.

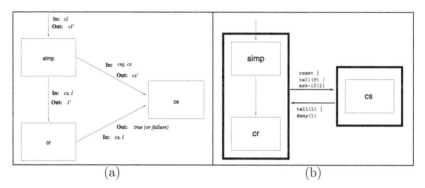

Figure 5. A control-level specification (a) and an agent-based architecture (b) of the case study.

- $l \prec l'$ asserts that the literal l is smaller than the literal l' w.r.t. a simplification ordering (see, e.g., [11] for a definition).

Let R be a set of conditional equations of the form $cnj \Rightarrow (s = t)$, where cnj is a set of literals intended conjunctively. The rules of the annotated reasoning theory are the following:[7]

$$\overline{cl \cup \{true\} \rightarrow_{\mathsf{simp}} \{true\}} \text{ cl-true} \qquad \overline{cl \cup \{false\} \rightarrow_{\mathsf{simp}} cl} \text{ cl-false}$$

$$\frac{\mathsf{cs\text{-}init}(cs_0) \quad \overline{cl} :: cs_0 \rightarrow_{\mathsf{cs}} cs \quad cs :: l \rightarrow_{\mathsf{cr}} l'}{cl \cup \{l\} \rightarrow_{\mathsf{simp}} cl \cup \{l'\}} \text{ cl-simp}$$

Rules `cl-true` and `cl-false` specify how to simplify a clause when $true$ and $false$ are in it, respectively. Rule `cl-simp` says that a literal l in a clause $cl \cup \{l\}$ can be replaced by a new literal l' obtained by rewriting l in context cs (premise $cs :: l \rightarrow_{\mathsf{cr}} l'$), where cs is obtained by extending the empty rewriting context cs_0 (premise $\mathsf{cs\text{-}init}(cs_0)$) with the negated literals in cl (premise $\overline{cl} :: cs_0 \rightarrow_{\mathsf{cs}} cs$).

$$\frac{\{\overline{l}\} :: cs \rightarrow_{\mathsf{cs}} cs' \quad \mathsf{cs\text{-}unsat}(cs')}{cs :: l \rightarrow_{\mathsf{cr}} true} \text{ cxt-ent}$$

$$\frac{cs :: cnj\sigma \rightarrow_{\mathsf{cr}} \emptyset \quad l[t\sigma]_u \prec l[s\sigma]_u}{cs :: l[s\sigma]_u \rightarrow_{\mathsf{cr}} l[t\sigma]_u} \text{ crew}$$

for each conditional rewrite rule $cnj \Rightarrow (s = t)$ in R. $cs :: cnj \rightarrow_{\mathsf{cr}} \emptyset$ abbreviates $cs :: l \rightarrow_{\mathsf{cr}} true$, for all $l \in cnj$ and $s[l\sigma]_u$ denotes the expression obtained from s by replacing the sub-expression at position u with

---

[7] If l is an atomic formula, then $\overline{l}$ stands for $\neg l$; if l is a negated atom of the form $\neg m$, then $\overline{l}$ abbreviates m. If cl is a set of literals, then $\overline{cl}$ abbreviates $\{\overline{l} \mid l \in cl\}$.

$l\sigma$. Rule cxt-ent asserts that a literal l can be rewritten to *true* in the rewrite context cs if the result of extending cs with the negation of l yields an inconsistent rewrite context (premise cs-unsat(cs')). Finally, rule crew says that the sub-expression $s\sigma$ at position u in the expression l can be rewritten to $t\sigma$ in the rewriting context cs if $cnj \Rightarrow (s = t) \in R$, σ is a ground substitution s.t. the literal $l[t\sigma]_u$ is $\prec$-smaller than $l[s\sigma]_u$ (premise $l[t\sigma]_u \prec l[s\sigma]_u$), and the instantiated conditions $cnj\sigma$ are entailed by the rewriting context cs (premise $cs :: cnj\sigma \to_{cr} \emptyset$).

4.2 An Agent-Oriented Architecture for CCR

As depicted in the schema of Figure 5(b), the agent-oriented architecture for CCR consists of two agents which interact by exchanging KQML messages. The agent on the left of Figure 5(b) encapsulates the top level simplification loop (simp) and the rewriter (cr) whereas the agent on the right encapsulates the decision procedure (cs).[8]

The protocol consists of the repeated application of the following pattern of interaction. Whenever simp tries to apply rule cl-simp it initiates the interaction with cs by issuing on the channel a message containing the reset performative. (The semantics of the reset performative is given in Figure 6.) This has the effect of initializing the constraint store of cs (cf. precondition cs-init(cs_0) in cl-simp). Next, simp sends cs the set of literals occurring in the context via the message tell($\overline{cl}$)[9] (cf. precondition $\overline{cl} :: cs_0 \to_{cs} cs$ in cl-simp) and finally it asks cr to rewrite the focus literal (cf. precondition $cs :: l \to_{cr} l'$). cr rewrites the input literal using the available rewrite rules and in doing this it may ask cs to determine whether the current focus literal is entailed by the context via the message ask(l). In reply to this request cs sends back a message of the form tell(l) or of the form deny(l). (Notice that cr may query cs also when it is trying to establish the conditions of a conditional rewrite rule.)

reset(A,B)		
Pre:		
Post:	$bel(B, c)$ iff $c \in cs_0$ where cs_0 is s.t. cs-init(cs_0)	

Figure 6. Semantics of the reset performative.

---

[8]For simplicity, we consider only two agents since we are interested in the interplay between the 'simplification' activity (namely clause simplification and rewriting) with the logical services provided by the decision procedure. However, the proposed methodology can easily be adapted to specify agent architectures with three or more agents.

[9]We omit the first two arguments of the performatives (namely the 'sender' and the 'recipient') whenever their identity can be inferred from the context.

In order to complete the description we must specify how the *bel* predicate is interpreted by the agents. Firstly we require that the decision procedure trusts the simplifier. This is formalized by the following axiom:

$$\forall c.(bel(\mathsf{simp}, c) \Rightarrow bel(\mathsf{cs}, c)).$$

This fact allows the decision procedure to extend its own knowledge base using the information issued by the simplifier via the `tell` performatives. Secondly we must specify the inference capability of the decision procedure. This is done by means of the following axiom:

$$\forall p.\forall cs.\forall cs'.((bel(\mathsf{cs}, p) \wedge bel(\mathsf{cs}, cs) \wedge bel(\mathsf{cs}, p :: cs \rightarrow_{\mathsf{cs}} cs')) \Rightarrow bel(\mathsf{cs}, cs'))$$

where $bel(\mathsf{cs}, cs)$ abbreviates $\bigwedge\{bel(\mathsf{cs}, c) : c \in cs\}$ and $bel(\mathsf{cs}, p :: cs \rightarrow_{\mathsf{cs}} cs'))$ states the provability of the sequent $p :: cs \rightarrow_{\mathsf{cs}} cs'$ in the annotated reasoning theory of Section 4.1.

5   Implementation and Interfaces

How can the ideas presented in this paper help with the implementation and management of MATHWEB? A general interaction protocol based on Internet standards transforms closed architectures like the MATHWEB-SB [14] and the LOGIC BROKER ARCHITECTURE [5] systems into an open MATHWEB architecture. Currently, the former uses the distributed programming features provided by the MOZART programming language [27] for communication, while the latter takes advantage of the communication functionality provided by CORBA [9]. As a consequence, mathematical services either have to be embedded into a MOZART agent wrapper, or have to implement a CORBA interface, or have to do both, if they want to communicate with services that are not present on both architectures.

In order to achieve a joint system that reuses much of the current functionality in a KQML-based architecture, it is sufficient to augment each of the systems above by an agent that does the KQML-communication and serves as a bridge. This agent is a KQML facilitator agent that listens to a given network port (by default the OPENMATH port 1473) and relays KQML messages to the other agents in its architecture (by MOZART or CORBA communication). This allows to reuse the existing implementations and internal communication among agents as well as providing a standardized interface to the Internet.

Note that this also allows other implementations than the ones listed above to participate in MATHWEB as long as they implement the same outward appearance (e.g. KQML, OMRS, and ⊙DOC). To simplify KQML message passing, we will for the moment identify agent names with transport addresses, since agent mobility seems not to be a problem in MathWeb. Agent names in MathWeb are quadruples of the form

$$\langle method \rangle : //\langle machine \rangle{:}\langle port \rangle \# \langle agent \rangle$$

that resemble URLs. For example the name of the broker agent would be `kqml-xml://mathweb.org:1473#broker`.

Since ⊙Doc and OpenMath use an Xml representation for mathematical objects, a first step for using Kqml in MathWeb is to supply an Xml encoding of Kqml. We have set up an Xml document type definition for Kqml (see `http://www.mathweb.org/omdoc/dtd/kqml.dtd`) based on the 1997 Kqml proposal by Finin and Labrou [25]. Finally, MathWeb-SB supports the commonly-used presentation-layer (see Figure 1) transport protocols `xml-rpc`, `http:get/put`, and sockets that can be used for `kqml-xml` message passing.

6   Conclusion

We have laid down the first ideas for implementing a communication protocol for reasoning services using Kqml and OMRS. The former provides the high-level performative- and message-layers, while the latter gives the specification infrastructure for determining (and for the agents to reason about) the meaning of interactions of reasoning services. Together with ⊙Doc [23] as a content language, this gives a suitable basis for communication of mathematical services in MathWeb [26].

The motivation for this paper and the general approach taken comes from our experience with the MathWeb-SB [14] and the Logic Broker Architecture [5] systems, and the perceived need for a standardized interaction layer. Both systems have a largely ad-hoc set of interaction primitives, following the needs of the growing systems. Conceptually it is clear, that all of these primitives can be mapped into Kqml, if we provide specification schemata for the internal states of reasoning systems, which we have started in this paper. The next step will be to implement the communication by `kqml-xml`. We have already implemented a `kqml-xml` interface for MathWeb. So it only remains to develop a `kqml-xml`-aware broker service and a Kqml-xml/Corba bridge.

For other MathWeb services, such as mathematical knowledge bases (e.g. the MBase system [21]) or symbolic computation systems, a corresponding ontology must still be developed, in order to access them via Kqml. For the former, the problem will be relatively simple as the Kqml views agents a virtual knowledge bases anyway, for the latter, a suitable variant of OMRS has been presented in [17].

References

[1] Alessandro Armando, Alessandro Coglio, Fausto Giunchiglia, and Silvio Ranise. The Control Layer in Open Mechanized Reasoning Systems: Annotations and Tactics. To appear in [20], 2000.

[2] Alessandro Armando and Silvio Ranise. Constraint Contextual Rewriting. In R. Caferra and G. Salzer, eds., *Proc. of the 2nd Intl. Workshop on First Order Theorem Proving (FTP'98)*, pages 65–75, 1998.

[3] Alessandro Armando and Silvio Ranise. From Integrated Reasoning Specialists to "Plug-and-Play" Reasoning Components. In Calmet and Plaza [7], pages 42–54.

[4] Alessandro Armando and Silvio Ranise. Termination of Constraint Contextual Rewriting. In H. Kirchner and Ch. Ringeissen, editors, *Proceedings of the 3rd International Workshop on Frontiers of Combining Systems, FroCoS'2000*, pages 47–61. Springer LNAI 1794, 2000.

[5] Alessandro Armando and Daniele Zini. Towards Interoperable Mechanized Reasoning Systems: the Logic Broker Architecture. In A. Poggi, editor, *to appear on the Proceedings of the AI*IA-TABOO Joint Workshop 'From Objects to Agents: Evolutionary Trends of Software Systems'*, Parma, Italy, May 29–30, 2000. (See also the article in this volume)

[6] P. G. Bertoli, J. Calmet, F. Giunchiglia, and K. Homann. Specification and Integration of Theorem Provers and Computer Algebra Systems. In Calmet and Plaza [7], pages 94–106.

[7] Jaques Calmet and Jan Plaza, editors. *Proceedings of the International Conference on Artificial Intelligence and Symbolic Computation (AISC-98)*, Springer LNAI 1476, 1998.

[8] Olga Caprotti and Arjeh M. Cohen. Draft of the Open Math standard. The Open Math Society, http://www.nag.co.uk/projects/OpenMath/omstd/, 1998.

[9] The Object Management Group. The Common Object Request Broker Architecture. http://www.corba.org/.

[10] Louise A. Dennis, Graham Collins, Michael Norrish, Richard Boulton, Konrad Slind, Graham Robinson, Mike Gordon, and Tom Melham.

The Prosper Toolkit. In *Proceedings of the Sixth International Conference on Tools and Algorithms for the Construction and Analysis of Systems, TACAS-2000*, Springer LNCS, 2000.

[11] N. Dershowitz and J.P. Jouannaud. Rewriting systems. In *Handbook of Theoretical Computer Science*, pages 243–320. Elsevier Publishers, Amsterdam, 1990.

[12] T. Finin and R. Fritzson. KQML — a Language and Protocol for Knowledge and Information Exchange. In *Proceedings of the 13th Intl. Distributed Artificial Intelligence Workshop*, pages 127–136, Seattle, WA, USA, 1994.

[13] Andreas Franke, Stephan M. Hess, Christoph G. Jung, Michael Kohlhase, and Volker Sorge. Agent-Oriented Integration of Distributed Mathematical Services. *Journal of Universal Computer Science*, 5:156–187, 1999.

[14] Andreas Franke and Michael Kohlhase. System Description: MATH-WEB, an Agent-Based Communication Layer for Distributed Automated Theorem Proving. In Harald Ganzinger, editor, *Proceedings of the 16th Conference on Automated Deduction*, pages 217–221. Springer LNAI 1632, 1999.

[15] D. Fuchs and J. Denzinger. Knowledge-Based Cooperation between Theorem Provers by Techs. Seki Report SR-97-11, Fachbereich Informatik, Universität Kaiserslautern, 1997.

[16] F. Giunchiglia, P. Pecchiari, and C. Talcott. Reasoning Theories: Towards an Architecture for Open Mechanized Reasoning Systems. Technical Report 9409-15, IRST, Trento, Italy, 1994.

[17] K. Homann. *Symbolisches Lösen mathematischer Probleme durch Kooperation algorithmischer und logischer Systeme*. PhD thesis, Unversität Karlsruhe, 1996. DISKI 152, Infix; St. Augustin.

[18] N. R. Jennings and M. Wooldridge. *Handbook of Agent Technology*, chapter Agent-Oriented Software Engineering. AAAI/MIT Press, 2000.

[19] Nicholas R. Jennings and Michael J. Wooldridge, editors. *Agent Technology : Foundations, Applications, and Markets*. Springer, 1998.

[20] Tudor Jebelean and Alessandro Armando, eds. *J. Symbolic Computation*. Special Issue on Integrated Symbolic Computation and Automated Deduction, 2000.

[21] M. Kohlhase and A. Franke. Mbase: Representing Knowledge and Context for the Integration of Mathematical Software Systems. to appear in [20], 2000.

[22] Michael Kohlhase. ©Doc: Towards an Internet Standard for the Administration, Distribution and Teaching of Mathematical Knowledge. In *Proceedings of AI and Symbolic Computation, AISC-2000*, Springer LNAI, 2000. in press.

[23] Michael Kohlhase. OMDoc: Towards an OPENMATH Representation of Mathematical Documents. Seki Report SR-00-02, FB Informatik, Universität des Saarlandes, 2000. http://www.mathweb.org/omdoc.

[24] Y. Labrou and T. Finin. A Semantics Approach for KQML—A General Purpose Communication Language for Software Agents. In *Third International Conference on Information and Knowledge Management (CIKM'94)*, November 1994.

[25] Yannis Labrou and Tim Finin. A Proposal for a New KQML Specification. Tech Report TR CS-97-03, University of Maryland, Baltimore County, 1997.

[26] The MathWeb group. MathWeb.org: Supporting Mathematics on the Web. http://www.mozart-oz.org/.

[27] The Oz group. The MOZART Programming System. http://www.mozart-oz.org/.

[28] M. Dewar ed. Special Issue on OPENMATH Bulletin of the ACM Special Interest Group on Symbolic and Algebraic Mathematic (SIGSAM), 2000, in press.

[29] C. Talcott. Reasoning Specialists Should Be Logical Services, Not Black Boxes. In *Workshop on Theory Reasoning in Automated Deduction (CADE12)*, pages 1–6, 1994.

Interfacing Computer Algebra and Deduction Systems via the Logic Broker Architecture

Alessandro Armando   Daniele Zini

Abstract. *In previous work we have introduced the* Logic Broker Archi-tecture, *a framework which provides the needed infrastructure for making mechanized reasoning systems interoperate. The architecture provides loca-tion transparency, a way to forward requests for logical services to appro-priate reasoning systems via a simple registration/subscription mechanism, and a translation mechanism which ensures the transparent and provably sound exchange of logical services. In this paper we show that the Logic Broker Architecture can be used as a firm basis for the integration of com-puter algebra systems and deduction systems. To substantiate our claim we discuss the coupling of the theorem prover* **RDL** *and the computer algebra system CoCoA via our current prototype implementation of the Logic Bro-ker Architecture which is based on CORBA and the* OPENMATH *standard.*

1 Introduction

In the recent years there has been a number of attempts to the integration of Deduction Systems (DSs) and Computer Algebra Systems (CASs) [7, 8, 9, 11, 16, 22, 24, 25]. The motivation stems from the consideration that such systems provide a high degree of sophistication in their respective domain but they perform poorly when applied outside the domain they have been designed for. In particular, CASs are usually good in performing heavy computation whereas DSs are not suited for such tasks; on the other hand DSs usually come equipped with expressive languages and this is a crucial feature in many practical applications.

It is thus expected that the integration of CASs and DSs will lead to a new generation of mechanized reasoning systems capable of combining the features of the two classes of systems. Unfortunately this is not an easy task. The main difficulty is that most of the existing CASs and DSs are conceived and built as stand-alone software systems mainly intended to be used by human users. Moreover, if the logical services provided by the component reasoning systems are not interfaced in a proper way, then the logical services provided by the compound systems may be unsound.

In [4] we have presented the *Logic Broker Architecture*, a framework which provides the needed infrastructure for making mechanized reasoning systems interoperate. The architecture provides location transparency, a way to forward requests for logical services to appropriate reasoning systems via a simple registration/subscription mechanism, and a translation mechanism which ensures the transparent and provably sound exchange of logical services. We believe the Logic Broker Architecture can be used as a firm basis for the integration of CASs and DSs. In this paper we substantiate this claim by discussing the coupling of the theorem prover **RDL** [3] with the computer algebra system CoCoA [12] via our current prototype implementation of the Logic Broker Architecture which is based on CORBA [20] and the OPENMATH standard [28].

The paper is organized in the following way. In Section 2 we present the Logic Broker Architecture. The prototype implementation of the Logic Broker Architecture is discussed in Section 3. In Section 4 we illustrate the case study. Finally, in Section 5, we compare with the related work.

2 An Overview of the Logic Broker Architecture

The basic schema of the Logic Broker Architecture is depicted in Figure 1, where a *client* reasoning system C gets access to the services provided by a *server* S via the Logic Broker LB. The rôle of the LB is threefold:

1. the LB provides location transparency for the reasoning systems by routing messages to systems without requiring senders to know the locations of the receivers,

2. the LB facilitates interoperation by providing a way for systems to discover which servers can handle which requests using a simple registration/subscription mechanism, and

3. the LB automatically translates the requests for logical services issued by the *client* into corresponding requests for the server.

Since location transparency is a domain-independent feature which can be readily obtained by using general purpose and well established frameworks such as CORBA, here we focus last two (domain specific) issues.

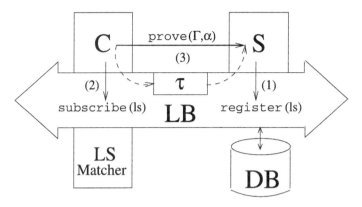

Figure 1. The Logic Broker Architecture.

2.1 The Registration/Subscription Mechanism

A reasoning system S can register to the LB as a *server* by sending the LB a message of the form `register(LS`$_s$`)`, where LS$_s$ is a specification of the logical service it is able and willing to provide (e.g. `prove`, `simplify`, `factorize`, ...). We assume that the specification of a logical service comes equipped with the specification of the associated consequence relation. A *consequence relation* [6], i.e. a pair of the form $(L, \vdash)$ where L is a set of sentences and $\vdash \subseteq \mathcal{P}(L) \times L$ (where $\mathcal{P}(L)$ denotes the power set of L) is a binary relation enjoying the following properties: (*Inclusion*) if $\alpha \in \Gamma$, then $\Gamma \vdash \alpha$; (*Monotonicity*) if $\Gamma \vdash \alpha$, then $\Gamma \cup \Delta \vdash \alpha$; (*Cut*) if $\Gamma \vdash \alpha$ and $\Delta \cup \{\alpha\} \vdash \beta$, then $\Gamma \cup \Delta \vdash \beta$—for all $\Gamma, \Delta \in \mathcal{P}(L)$ and $\alpha, \beta \in L$.

When LB receives a message of the form `register(LS`$_s$`)`, it stores the pair $\langle S, LS_s \rangle$ in a database of registered logical services for later use. Dually, a reasoning system can subscribe to the LB as a *client* by issuing the LB a message of the form `subscribe(LS`$_c$`)`, where LS$_c$ is a specification of the requested service. Upon receipt of such a message, the LB searches the database of registered services for a pair $\langle S, LS_s \rangle$ such that the specification of the logical service LS$_s$ has the same interface of the required one, and checks whether there exists a morphism ϕ from $(L_s, \vdash_s)$ into $(L_c, \vdash_c)$, i.e. a function $\phi : L_s \to L_c$ such that $\Gamma \vdash_s \alpha$ implies $\phi(\Gamma) \vdash_c \phi(\alpha)$ for all $\Gamma \in \mathcal{P}(L_s)$ and $\alpha \in L_s$. ($\phi(\Gamma)$ abbreviates $\{\phi(\gamma) : \gamma \in \Gamma\}$.) It readily follows from the definition that if there exists a morphism ϕ from $(L_s, \vdash_s)$ into $(L_c, \vdash_c)$ then a problem of the form $\Gamma_c \vdash_c \alpha_c$ for some $\Gamma_c \in \mathcal{P}(L_c)$ and $\alpha_c \in L_c$ arising at the *client* side can always be reduced to asking the server whether $\Gamma_s \vdash_s \alpha_s$ for any $\Gamma_s \in \mathcal{P}(L_s)$ and $\alpha_s \in L_s$ such that $\phi(\Gamma_s) = \Gamma_c$ and $\phi(\alpha_s) = \alpha_c$.

The task of determining the existence of a morphism between from $(L_s, \vdash_s)$ into $(L_c, \vdash_c)$ is carried out by a specialized reasoning module called the *Logical Service Matcher* (*LS Matcher* for short). Notice that the problem is not decidable in the general case and therefore user intervention may be necessary here.

If the above steps terminate successfully then the LB establishes a connection between the *client* and the *server*.

2.2   The Translation Mechanism

Let ϕ be the morphism determined by the LS Matcher as explained in Section 2.1. The LB defines a function $\tau : L_c \to L_s$ which translates a formula α_c of the *client* into a formula α_s of the *server* such that $\phi(\alpha_s) = \alpha_c$ if α_c is in the range of ϕ and it is undefined otherwise.

Let `prove` be a logical service obtained by the *client* as a reply to a `subscribe` request sent to the LB; whenever the *client* issues a message of the form `prove`($\Gamma \vdash_c \alpha$) to the *server*, then the LB transparently carries out the following steps:

1. Γ and α are translated into $\tau(\Gamma)$ and $\tau(\alpha)$ respectively,

2. the message `prove`($\tau(\Gamma) \vdash_s \tau(\alpha)$) is delivered to the *server*, and

3. the answer returned by the *server* is delivered back to the *client*.

It readily follows from the definition of morphism that whenever the *client* gets a positive reply to a message of the form `prove`($\Gamma \vdash_c \alpha$), then $\Gamma \vdash_c \alpha$ holds. This is an important result which ensures the logical soundness of the interaction between the *client* and the *server provided that the logical services provided by the server comply with the specification given as argument of the corresponding* `register` *messages*. The problem of ensuring that the logical services meet with their specifications is out of the scope of the LB. It is therefore up to the *client* to trust or not the *server* and in the latter case to undertake appropriate actions as done, e.g., in [23].

3   The Current Implementation of the Logic Broker Architecture

We have built a first running prototype of the Logic Broker Architecture using CORBA and the OPENMATH standard. CORBA is an industrial standard which provides interoperability and therefore its the candidate of choice for many reasons.

CORBA. The advantage of using CORBA as a platform for the development of the Logic Broker Architecture is threefold. First of all CORBA greatly simplifies the activity of combining reasoning systems since it is specifically designed to reduce to a minimum the burden associated with the activity of interfacing software systems. Secondly, CORBA gives us location transparency for free and provides us with general and robust solutions to the problems associated with distributed memory management and event handling. Last but not least, CORBA is an industrial standard and many ORB implementations (both free and commercial) are publicly available.

CORBA is an architecture based on the *Object Oriented* paradigm, and built around three key notions: the OMG Interface Definition Language (IDL), the Object Request Broker (ORB), and the standard protocol IIOP for inter-ORB communication. Each object in a CORBA application has an interface defined in IDL. The interface definition of the of the LB is given in Figure 2. It states that *(i)* the `Register` functionality takes as input a server s and the specification of a logical service to be registered `ls` and that *(ii)* the `Subscribe` functionality takes as input the specification of a logical service and returns a logical service.

The IDL interface definition does not commit to a specific programming language, but *maps* to most programming languages via a set of OMG standards. When a client needs to invoke an operation on a server the ORB transparently routes this request to the server.[1] This activity is supported by software interfaces called *stub* and *skeleton* at the client side and at the server side, respectively. The *stub* and the *skeleton* are generated *automatically* by IDL compilers in different programming languages. For example if a server is implemented in C++ then the *skeleton* can be generated in C++, whereas if a client is implemented in Java then the *stub* can be generated in Java.

```
interface LogicBroker
{
   void Register(in LSServer s, in LSSpec ls);
   LogicService Subscribe(in LSSpec ls);
};
```

Figure 2. The IDL of the Logic Broker.

---

[1] The ORB uses the standard IIOP protocol to transparently provide location transparency over the Internet.

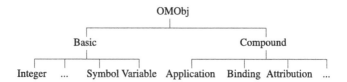

Figure 3. The hierarchy of the OPENMATH Objects.

OPENMATH. We recall from the previous sections that the main difficulty in matching logical services is that of finding a morphism between the associated consequence relations. Unfortunately this problem is undecidable and therefore a solution which is both general and fully automatic is not possible. The solution we adopted is based on the an emerging standard for representing mathematical knowledge called OPENMATH [14]. In particular we exploited the following features of the OPENMATH standard:

- OPENMATH provides a common representation for mathematical expressions via OPENMATH objects. OPENMATH *objects* are (abstractly defined) mathematical expressions built out of primitive constructs for atomic entities (such as symbols, variables, and numbers) and constructs for compound objects such as applications, bindings, and attributions (i.e. annotations) as illustrated in Figure 3.

- OPENMATH provides a standard ontology for mathematical domains based on the notion of *content dictionary* (CD for short). CDs specify the semantic of OPENMATH symbols by providing the following information: the *name* of the symbol, the *description* in natural language, an *example*, the *signature*, a *commented mathematical property* (CMP for short), and a *formal mathematical property* (FMP for short). The first two items are strings of characters whereas the last four are OPENMATH objects. While both the *example* and the CMP specify the semantic at an informal level, the *signature* and the FMP provide the formal semantics. The entry of the OPENMATH CD `arith1` for the OPENMATH Symbol `plus` is given in Figure 4.

The choice of the OPENMATH standard[2] simplified considerably the development of our prototype. The main implementation effort amounted to

---

[2] We adopt the OPENMATH standard at the abstract level since we do not commit to specialized encodings for OPENMATH Objects and Content Dictionary. In our framework CORBA frees the programmer from the burden of dealing with the actual encoding of the information exchanged.

Name: plus

Description: The n-ary commutative function plus.

Commented Mathematical Property (CMP): a + b = b + a

Formal Mathematical Property (FMP):
forall [a b] . (eq (plus (a, b), plus (b, a)))

Figure 4. OPENMATH Symbol plus as defined in the CD arith1.

the definition of the interfaces for OPENMATH objects, the Logical Services, the Reasoning Systems (both *client* and *server*), the Logic Broker and the LS Matcher using the OMG Interface Definition Language (IDL for short). The IDL definition for the Logical Services is given in Figure 5. It states that a logical service provides a single functionality that takes an OPEN-MATH object as input and returns a new OPENMATH object. For instance the logical service of a simplifier takes (an OPENMATH object representing) an expression as input and returns (an OPENMATH object representing) the simplified expression.

In order to connect to reasoning systems via our prototype implementation of the Logic Broker Architecture, the reasoning systems are required to translate mathematical statements expressed in their respective logical language into/from equivalent OPENMATH objects so to preserve the semantics encoded in the Content Dictionaries. Following OPENMATH, we call *phrasebooks* the procedures in charge of the translations.[3] As in the general case, the big advantage of using a standard communication language is that a single translation from each private language to the stan-

```
interface LogicalService
{
OMObject execute(in OMObject omobj);
};
```

Figure 5. The IDL interface for Logical Services.

---

[3]Notice that in the OPENMATH standard a phrasebook has to deal with a specific encoding such as XML or Binary encodings but in our architecture the OPENMATH Objects are encoding independent via the IDL interface.

dard is necessary as opposed to the general case in which a translation for each pair of private languages is needed.

Technically speaking phrasebooks must be morphisms between the consequence relations mechanized by the reasoning systems and that encoded in the OPENMATH Content Dictionaries. However the current implementation of the Logic Broker Architecture does not provide support to this activity and therefore this task (and responsibility) is left to the implementor of the phrasebook. In the future work we plan to incorporate into our implementation of the Logic Broker Architecture a reasoning system which should play the role of the LS Matcher. The LS Matcher will allow us formalize the consequence relations of the reasoning systems connected to the Logic Broker Architecture and reason about the relationship with the information specified in the FMPs stored in the OPENMATH Content Dictionaries. In particular the LS Matcher will provide support to the activity of synthesising the phrasebooks and certifying that they are indeed morphisms. We will return on this issue in Section 5.

4 Case Study: Interfacing CoCoA and **RDL** via the Logic Broker Architecture

RDL [3] is an automatic theorem prover which combines conditional rewriting with decision procedures for integer linear arithmetic and ground equality. [2] describes an extension mechanism which enables the decision procedure for integer linear arithmetics incorporated into **RDL** to tackle nonlinear problems of significant complexity. To illustrate let us consider the problem (from [10]) of showing the unsatisfiability of the formula

$$
\begin{aligned}
ms(c) + ms(a)^2 + ms(b)^2 \geq \\
ms(c) + ms(b)^2 + 2 * ms(a)^2 * ms(b) + ms(a)^4
\end{aligned}
\tag{11}
$$

using the following fact as lemma

$$
\forall X.(0 < ms(X))
\tag{12}
$$

The decision procedure simplifies (11) into

$$
ms(a)^4 + 2 * ms(a)^2 * ms(b) - ms(a)^2 \leq 0
\tag{13}
$$

and then it gets stuck. The extension mechanism comes into play by first factorizing (13) into

$$
ms(a)^2 * (ms(a)^2 + 2 * ms(b) - 1) \leq 0
\tag{14}
$$

and then recognizing that (14) is an instance of a *hyperbolic inequality*, i.e. a formula of the form $S*T \leq k$ where k is a numeric constant. As shown in [27] a hyperbolic inequality can always be transformed into an equivalent formula in disjunctive normal form built out of inequalities (linear in S and T) whose form and size depend from the actual value of k. If $k = 0$—as it is our example—$S*T \leq k$ is trivially equivalent to $(S \leq 0 \wedge T \geq 0) \vee (S \geq 0 \wedge T \leq 0)$, i.e. the following fact holds:

$$S * T \leq 0 \leftrightarrow ((S \leq 0 \wedge T \geq 0) \vee (S \geq 0 \wedge T \leq 0)). \tag{15}$$

(15) can be weakened to

$$S > 0 \rightarrow (S * T \leq 0 \leftrightarrow ((S \leq 0 \wedge T \geq 0) \vee (S \geq 0 \wedge T \leq 0))),$$

which is then simplified (by simple arithmetic and propositional reasoning) to:

$$S > 0 \rightarrow (S * T \leq 0 \leftrightarrow T \leq 0). \tag{16}$$

The instance of obtained by replacing S and T with $ms(a)^2$ and $ms(a)^2 + 2 * ms(b) - 1$ respectively allows us to reduce the problem of the unsatisfiability of (14) to that of

$$ms(a)^2 + 2 * ms(b) - 1 \leq 0 \tag{17}$$

under the proviso that the following fact holds:

$$ms(a)^2 > 0. \tag{18}$$

Both (17) and (18) are then readily solved by recursively applying the extended procedure and using (12). (See [2] for the details.)

A crucial step of the extension mechanism is the factorization of arithmetic expressions. In the above example the factorization is fairly simple, but in the general case things are much more involved. Fortunately factorization is one of the computational services computer algebra systems are particularly good at. Therefore we decided to couple **RDL** with CoCoA, a computer algebra system capable of a variety of operations on multivariate polynomial rings, via the Logic Broker Architecture as depicted in Figure 6 so that the extension mechanism of **RDL** can invoke CoCoA whenever a factorization is necessary. To give a concrete feel of the actual practical usage of the Logic Broker Architecture here we illustrate what we have carried out to achieve the coupling.

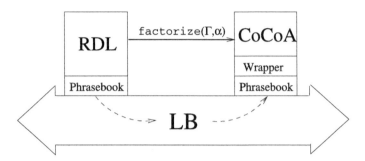

Figure 6. The RDL-CoCoA combination via the Logic Broker Architecture.

Server Side. We first interfaced CoCoA to the Logic Broker Architecture by means of the following steps:

1. **Wrapping up CoCoA.** This activity amounted to implementing a few C++ functions capable of managing and communicating with a CoCoA process. In particular the wrapper gives access to the `factor` function of CoCoA.

2. **Building an** OPENMATH **phrasebook for CoCoA.** We implemented a simple phrasebook capable of translating (a suitable subset of) OPENMATH Objects into corresponding CoCoA expressions and vice-versa. For instance, the phrasebook translates the OPENMATH Object representing the expression $ms(a)^4 + 2 * ms(a)^2 * ms(b) - ms(a)^2$ into the CoCoA expression `v[0]^4+2*v[0]^2*v[1]-v[0]^2`. (`v[0]` and `v[1]` are indeterminates in CoCoA if—as we assume here—the "current" ring is defined to be `Q[v[0..1]]`. See [12] for the details.)

3. **Getting the Skeleton from the IDL.** We compiled the IDL interface shown in Figure 5 into the skeleton in the programming language of choice. (We chose C++ and used ORBACUS [21] as ORB implementation in our case study.) The IDL compiler automatically generates the class `LogicalService_skel` along with all the machinery to handle clients' requests coming through the ORB.

4. **Filling in the Skeleton.** This is the only implementational effort required and it simply amounts to linking the functionalities specified in the skeleton to those provided by the wrapper. In our case study we provided an implementation for the interface of Figure 5 which first issue CoCoA the command `Use R::=Q[v[0..n]];` (n is the number

of different multiplicands in the polynomial to be factorized) and then it issues the command `Factor(p)` (p if the CoCoA representation of the input polynomial) and returns the result computed by CoCoA.

Client Side. Here are the steps we carried out in order to attach **RDL** to the Logic Broker Architecture.

1. **Getting the Stub from the IDL.** We compiled the IDL interface shown in Figure 5 into the stub in the programming language of choice. (We choose C++[4]). The IDL compiler automatically generates the class `LogicalService` whose methods transparently invoke the server via the ORB.

2. **Interfacing RDL.** As **RDL** is implemented in Prolog and since compilers from IDL to Prolog are not yet available we linked the functionalities provided by the stub to a corresponding Prolog predicate via the SICStus Prolog Foreign Language Interface (FLI).

3. **Building an** OPENMATH **phrasebook for RDL.** We implemented a simple phrasebook capable of translating (a suitable subset of) OPEN-MATH Objects into corresponding expressions of **RDL** and vice-versa. For instance, the phrasebook translates the OPENMATH Object representing the expression $ms(a)^4 + 2 * ms(a)^2 * ms(b) - ms(a)^2$ into the **RDL** expression `pow(ms(a),4)+2*pow(ms(a),2)*ms(b)-pow(ms(a),2)`.

To illustrate, let us consider the situation in which **RDL** must factorize the polynomial $ms(a)^4 + 2 * ms(a)^2 * ms(b) - ms(a)^2$. To this end **RDL** invokes the predicate `factorize(pow(ms(a),4)+2*pow(ms(a),2)*ms(b)-pow(ms(a),2),P)`. The SICStus Prolog FLI first invokes the phrasebook in order to get the OPENMATH Object corresponding to `pow(ms(a),4)+2*pow(ms(a),2)*ms(b)-pow(ms(a),2)`, say p, then it invokes the method `LogicalService->execute(p)`. The LB reacts to the invocation of this method at the client side by issuing the call to the server side and awaits for the reply. The result of the factorization is then returned to the client side as an OPENMATH Object which is then turned into an **RDL** expression by the phrasebook.

Notice that the management of the communication is transparently provided by our implementation of the Logic Broker Architecture. In particular it is worth pointing out that the OPENMATH Objects generated

---

[4]Of course the programming language at the *client* side can be different from that at the *server* side. In our case study we just found convenient to use C++ at both sides.

by the phrasebook are not created locally (to the phrasebook) but re-
motely. The underlying ORB takes care of the distributed management
(namely, creation and deallocation) of the OPENMATH Objects. This is
a very important feature which dramatically simplifies the development of
distributed systems in general.

5 Related Work

The OPENMATH Project is obviously very relevant to ours, but—as we
have already pointed out in previous sections—the objectives are largely
complementary. While OPENMATH focuses on the definition of a language
for representing and communicating mathematics, the Logic Broker Architecture provides the infrastructure for making reasoning systems cooperate
in a principled and sound way.

The PROSPER (Proof and Specification Assisted Designed Environment) Project aims at "[...] researching and developing a toolkit that
allows an expert to easily and flexibly assemble *proof engines* to provide
embedded formal reasoning support inside applications." [13]. The toolkit
allows the user to assemble a custom built verification engine out of the
HOL theorem prover possibly empowered by 'plugins' formed from existing, off-the-shelf, tools. Communication between the theorem prover and
the application as well as that occurring between the theorem prover and
the plugins must comply with the PROSPER Integration Interface. We believe that our prototype implementation of the Logic Broker Architecture
already enjoys many features of the PROSPER Toolkit and can be used for
the same purpose. The main advantage is that—being based on CORBA—
our prototype is open and extensible whereas the PROSPER toolkit is based
on a proprietary protocol.

MathWeb [17] is a distributed network architecture for automated and
interactive theorem proving aiming at supporting modularization, interoperability, robustness, and scalability of mathematical software systems. The
main features of MathWeb are largely complementary to those of the Logic
Broker Architecture. In MathWeb the activity of combining logical services is simplified considerably thanks to the adoption of the Agent-Orient
Programming paradigm [18]. On the other hand, MathWeb makes no pro-
vision to ensure the soundness of the combination of the logical services,
whereas this is one of the main objectives of the Logic Broker Architecture.
Furthermore, MathWeb agents employ a proprietary protocol based on an
XML encoding of (a subset of) KQML performatives. In our prototype
implementation of the Logic Broker Architecture the integration of new

reasoning systems is done at a higher level of abstraction as CORBA hides the details of the actual encoding.

The Open Mechanized Reasoning Systems (OMRS for short) project [19, 1] aims at the definition of a specification framework for specifying logical services. Specifications play a fundamental rôle when automatic and/or provably sound integrations are at stake. We envisage that the OMRS specification framework will play a pivotal rôle in the specification of logical services and their interplay in the Logic Broker Architecture.[5]

6 Conclusions and Future Work

We have presented the *Logic Broker Architecture*, a framework which provides the middle-ware for making mechanized reasoning systems interoperate. The architecture provides location transparency, a common syntax for representing mathematical information, and a way to forward requests for logical services to appropriate reasoning systems via a simple registration/subscription mechanism.

In the future work we will address the problem of adding a LS Matcher to our prototype implementation of the Logic Broker Architecture. A few existing deduction systems already exist which can be used for this purpose. IMPS [15] is a theorem prover which supports the structured presentation of theories and has heuristics for finding morphisms between them. [5] introduces a data structure, called *development graph*, devoted to the modular representation and management of specifications (abstractly characterized as consequence relations). As development graphs are independent from the specific specification language, they provide a more general framework than that provided by IMPS which is confined to the more specific notion of theory morphism.

References

[1] A. Armando, A. Coglio, and F. Giunchiglia. The Control Component of Open Mechanized Reasoning Systems. *Electronic Notes in Theoretical Computer Science*, 23(3):3–20, 1999.

[2] A. Armando and S. Ranise. A Practical Extension Mechanism for Decision Procedures. In *In Proc. of the 4th Workshop on Tools for*

---

[5]The current prototype implementation of the Logic Broker Architecture relies on a very simple language to this end.

System Design and Verification (FM-TOOLS'2000, Reisensburg Castle near Ulm, Germany, July 2000. Available at `http://www.mrg.dist.unige.it/~silvio/docs/fmtools00.ps.gz`.

[3] A. Armando and S. Ranise. System Description: **RDL** – Rewrite and Decision procedure Laboratory, 2000. DIST Technical Report – University of Genova.

[4] A. Armando and D. Zini. Towards Interoperable Mechanized Reasoning Systems: the Logic Broker Architecture. In *Proc. of the AI*IA-TABOO Joint Workshop 'Dagli Oggetti agli Agenti*, Parma, Italia, May 29-30 2000.

[5] Serge Autexier, Dieter Hutter, Heiko Mantel, and Axel Schairer. Towards an evolutionary formal software-development using CASL. In *Proc. Workshop on Algebraic Development Techniques WADT-99*, volume LNCS 1827, 1999.

[6] A. Avron. Simple consequence relations. LFCS Report Series, Laboratory for the Foundations of Computer Science, Computer Science Department, University of Edinburgh, 1987.

[7] C. Ballarin, K. Homann, and J. Calmet. Theorems and Algorithms: An Interface between Isabelle and Maple. In *Proc. of International Symposium on Symbolic and Algebraic Computation (IS-SAC'95)*, pages 150–157, Berkeley, CA, USA, 1995.

[8] B. Buchberger and T. Jebelean, editors. *Proceedings of the Second International Theorema Workshop*, RISC-Reports series No. 98-10. RISC-Hagenberg, Austria, 29-30 June, 1998.

[9] B. Buchberger, T. Jebelean, F. Kriftner, M. Marin, E. Tomuţa, and D. Văsaru. An Overview of the *Theorema* Project. In *Proc. ISSAC-97* Maui, HI, USA, July 21-23, 1997.

[10] R.S. Boyer and J S. Moore. Integrating Decision Procedures into Heuristic Theorem Provers: A Case Study of Linear Arithmetic. *Machine Intelligence*, 11:83–124, 1988.

[11] B. Buchberger, D. Văsaru, and T. Ida, editors. *Proceedings of the First International Theorema Workshop*, RISC-Reports series No. 97-20. RISC-Hagenberg, Austria, 9-10 June, 1997.

[12] A. Capani, G. Niesi, and L. Robbiano. *CoCoA, a system for doing Computations in Commutative Algebra*. Available via anonymous ftp from cocoa.dima.unige.it, 3.7 edition, 1999.

[13] L. A. Dennis, G. Collins, M. Norrish, R. Boulton, K. Slind, G. Robinson, M. Gordon, and T. Melham. The PROSPER Toolkit. In *Proc. of the 6th International Conference on Tools and Algorithms for the Construction and Analysis of Systems*, volume 1785, Berlin, Germany, April 2000.

[14] S. Dalmas, M. Gaëtano, and S. Watt. An OpenMath 1.0 implementation. In *Proc. ISSAC-97* Maui, HI, USA, July 21-23, 1997.

[15] William M. Farmer, Joshua D. Guttman, and F. Javier Thayer. Little theories. In *Proc. of the 11th International Conference on Automated Deduction (CADE-11)*, volume 607 of *LNAI*, pages 567–581, Saratoga Springs, NY, June 1992. Springer.

[16] A. Franke, S.M. Hess, C.G. Jung, M. Kohlhase, and V. Sorge. Agent-oriented integration of distributed mathematical services. *Journal of Universal Computer Science*, 5:156–187, 1999.

[17] Andreas Franke and Michael Kohlhase. System description: MATH-WEB, an agent-based communication layer for distributed automated theorem proving. In *Proc. CADE-16*, volume 1632 of *LNAI*, pages 217–221, Berlin, July 7–10, 1999. Springer-Verlag.

[18] Michael R. Genesereth and Steven P. Ketchpel. Software agents. *Communications of the ACM*, 37(7):48–53, July 1994.

[19] F. Giunchiglia, P. Pecchiari, and C. Talcott. Reasoning Theories: Towards an Architecture for Open Mechanized Reasoning Systems. Tech. Rep. 9409-15, IRST, Trento, Italy, 1994. Short version in Proc. of FroCoS'96, Munich, Germany, March 1996.

[20] Object Management Group. The Common Object Request Broker: Architecture and Specification. Technical report, OMG, July 1995. Rev. 2.0.

[21] Object Oriented Concepts Inc. ORBacus 3.0. Users manual 1.0 http://www.ooc.com, 1998.

[22] J. Harrison and L. Thery. Extending the HOL theorem prover with a computer algebra system to reason about the reals. In *Proc. International Workshop on Higher Order Logic Theorem Proving and its Applications*, volume 780 of LNCS, pages 174–185, Vancouver, Canada, August 1993.

[23] J. Harrison and L. Thery. A sceptic's approach to combining HOL and Maple. *J. of Automated Reasoning*, 21:279–294, 1998.

[24] M. Kerber, M. Kohlhase, and V. Sorge. Integrating Computer Algebra with Proof Planning. In *Proc. of DISCO '96*, volume 1128 of *LNCS*, Karlsruhe, Germany, september 18-20 1996.

[25] M. Kerber, M. Kohlhase, and V. Sorge. Integrating Computer Algebra Into Proof Planning. *J. of Automated Reasoning*, 21(3):327–355, 1998.

[26] Michael Kohlhase. OMDOC: Towards an OPENMATH representation of mathematical documents. Seki Report SR-00-02, Fachbereich Informatik, Universität des Saarlandes, 2000. http://www.mathweb.org/ilo/omdoc.

[27] V. Maslov and W. Pugh. Simplifying Polynomial Constraints Over Integers to Make Dependence Analysis More Precise. Technical Report CS-TR-3109.1, Dept. of Computer Science, University of Maryland, 1994.

[28] D. P. Carlisle O. Caprotti and A. M. Cohen. The OpenMath Standard. Technical Report D1.3.3a (Draft) (Public), The OpenMath Esprit Consortium, August 1999.

Development of the Theory of Continuous Lattices in MIZAR

Grzegorz Bancerek

Abstract. *This paper reports on* MIZAR *formalization of the theory of continuous lattices included in the* A Compendium of Continuous Lattices, *[7].* MIZAR *formalization means a formalization of theorems, definitions, and proofs in the* MIZAR *language such that it is accepted by the* MIZAR *system. This effort was originally motivated by the question whether the* MIZAR *system is sufficiently developed as to allow expressing advanced mathematics. The current state of the formalization, which includes 49* MIZAR *articles written by 14 authors, suggests that the answer is positive. The work of the team of authors in cooperation with the Library Committee[1] and system designers resulted in improvements of the system towards a more convenient technology for doing mechanically checked mathematics. It revealed, also, that the substantial element of the convenience is the incorporation of computer algebra into* MIZAR *system.*

1 Introduction

To formalize means to investigate some mathematical theory rigorously, obeying fixed rules of formulating, defining, proving, and reasoning. MIZAR formalization admits some variety of expression but the required rigor assures that the result of the formalization, that is a text in the context of the MIZAR system, has unique meaning. When formalizing a theory we introduce definitions, lemmas, and theorems with the hope that they will be useful for future developments. This is the essential idea behind developing the MIZAR data base.

---

[1] The Library Committee of the Association of Mizar Users maintains the Mizar Mathematical Library and coordinates activities concerning improvement of the library.

MIZAR² has been designed by Andrzej Trybulec and developed by a team under his leadership. The system includes a language, software tools, a library, and a hyperlinked journal.

The MIZAR language is an attempt to approximate mathematical vernacular in a formal language. Reserved words form a subset of English words which are used in regular mathematical papers with the same meaning. The logic of MIZAR is classical, the proofs are written in the Fitch-Jaśkowski style, see [9]. Definitions allow to introduce type, term, and formula constructors and require proving of correctness conditions. A proof consists of a sequence of steps, each step justified by facts proved in earlier steps, lemmas, theorems and/or schemes. Schemes are second order theorems which may be used to formulate e.g. induction. Multi prefixed structures allow to introduce algebraic concepts, for example topological groups which are both groups and topological spaces. More detailed description of MIZAR system can be found in [6], [18], and, also, in [13].

MIZAR software includes tools supporting some typical tasks when doing mathematics:

- development and management of knowledge base,

- verification of logical correctness,

- elements of generalization, simplification, readability enhancement,

- presentation using TeX and HTML.

The Mizar Mathematical Library (MML) is a collection of texts written in MIZAR language called *Mizar articles*. The MML is based on the Tarski-Grothendieck set theory. As of March 2000 there were 633 articles collected. They included 29,514 theorems, 5,389 definitions and redefinitions, and 317,427 references to external theorems (i.e. in other articles).

Mizar articles are automatically translated into English and published in *Formalized Mathematics*. The electronic version, *Journal of Formalized Mathematics*, http://www.mizar.org/JFM/ includes hyper-links to definitions which substantially help in using the MML. More details may be found on MIZAR web pages at http://www.mizar.org/.

MIZAR might be considered as an Esperanto for mathematics and there are some similarities between both languages. Esperanto was developed in Białystok by L. Zamenhoff and MIZAR is being developed in Białystok. Esperanto is an artificial international language with words taken from several national languages and uses a quite regular grammar. MIZAR may

---

²Mizar is a star; ζ in Ursa Major.

be considered as an attempt to standardize the language of mathematics. There is a lot of translations of books into Esperanto and it is possible to learn the language by reading those translations. There is a large collection of MIZAR articles and it is possible to learn MIZAR (and mathematics) by reading them.

2 Goal and Motivations

At the QED[3] Workshop II, Warsaw 1995, the following question was raised:

> Can we do formalization of advanced mathematics like this included in regular mathematical books in the current proof-checking systems?

In trying to answer this question we have decided to put MIZAR into a serious test. We have chosen *A Compendium of Continuous Lattices* [7] to be formalized in its entirety. The theory of continuous lattices presented in [7] is mathematically advanced. It involves a variety of areas of mathematics: computation, topology, analysis, algebra, category theory, and logic. Also, it is a relatively recent and a well-established field. The choice turned out to be a lucky one. The compendium is very rigorous which made the formalization comparatively easy; also, some initial fragments of the theory of lattices had been already developed in MIZAR.

In the past, there were some attempts to formalize entire mathematical books in computerized proof-checking systems. In the 1970's, Jutting [17] formalized Landau's *Grundlagen* [12] in AUTOMATH. Another attempt was the formalization of 2 chapters of *Theoretical Arithmetic* by Grzegorczyk, [8], in the 1980's. It was done by A. Trybulec's team in MIZAR 2, which was not equipped with the library. In MIZAR 2, each text was processed separately from other texts. All background knowledge needed to write a text was put without proofs in a preliminary section.

In 1989 we started to collect all MIZAR texts and on this basis develop and maintain the Mizar Mathematical Library. Each new MIZAR article can be submitted to the MML if it is accepted by the MIZAR verifier and refers only to articles already included in the MML. At the start, the basis of the MML was formed by two axiomatic articles:

- *Tarski Grothendieck set theory* [15],

- *Built-in concepts* [14], including strong arithmetic of real numbers.

---

[3] http://www-unix.mcs.anl.gov/qed/

The latter became a regular MIZAR article in 1998 when the construction of real numbers was completed.

The experiment with formalization of an entire book has many aspects and we expected to get answers to the following:

- Is the MIZAR language sufficiently expressive to formulate definitions, theorems, and proofs contained in [7]?

- Is MML rich enough to even start the formalization? Did MML cover the knowledge assumed in the compendium as background?

- Can the different concepts already defined in independent articles in MML be used together?

- Is the MIZAR software capable of handling this amount of material?

We hoped that running such an experiment, irrespective of the answers to the above questions, would lead to an improvement of MIZAR.

3   Teamwork

The work performed on the formalization is a result of a team effort by the researchers and students of the Institute of Mathematics and Institute of Computer Science, University of Białystok. MIZAR articles written in the project have been authored by: Czesław Byliński, Adam Grabowski, Ewa Grądzka, Jarosław Gryko, Artur Korniłowicz, Beata Madras, Agnieszka J. Marasik, Robert Milewski, Adam Naumowicz, Piotr Rudnicki (University of Alberta, Canada), Bartłomiej M. Skorulski, Andrzej Trybulec, Mariusz Żynel, and Grzegorz Bancerek.

In the summer of 1995, we started a seminar devoted to the theory of continuous lattices following [7] and [10]. In the spring of 1996, the final decision on formalization of [7] was made. Parts of the first two chapters, *O. A Primer of Continuous Lattices* and *I. Lattice Theory of Continuous Lattices*, were assigned to individual team members for formalization. We adopted the following rules:

- formalization is divided into two series of MIZAR articles with the identifiers[4]:

    - YELLOW - articles bridging the MML and the knowledge assumed in the compendium,

---

[4]Each MIZAR article, besides regular title, has an identifier which is used when referring to it.

- WAYBEL[5] - articles formalizing the main course of the compendium,

- no formalization of examples unless necessary,

- the formalization is as close to [7] as possible but taking into account some MIZAR peculiarities such as built-in concepts and mechanisms, possibility of automatic generalization, reuse of the MML, etc.

- the formalization should be more general than the theory in the compendium, e.g. we follow hints at generalization included in exercises (see 1.26 on page 52 of [7] and compare pages 38–42 with [4]).

Because of the number of people involved, the work was organized in a different way than the usual sequential contributions of articles to MML. Usually, an author writes an article and the article is not available to other authors until it is submitted to the MML. We wanted to formalize different parts of the book simultaneously as sequential development would be too slow. We decided to maintain a local library of YELLOW and WAYBEL series with completed and non-completed articles. This allowed for some parallelism in writing articles. The articles from local library were tested by later ones that used them and if there was a need they were revised. After some time they were presented on a seminar to discus possible generalization and, finally, submitted to the MML.

The size of the YELLOW series (17 articles of 49) indicates that MML was almost ready for the formalization. However, the following topics had to be developed:

- upper and lower bounds, suprema and infima (YELLOW_0),

- poset under inclusion (YELLOW_1),

- lattice of ideals (YELLOW_2),

- complete lattices (YELLOW_0 and YELLOW_2),

- Cartesian product of posets and lattices (YELLOW_1, YELLOW_3, and YELLOW10),

- lattice operations on subsets of a poset (YELLOW_4),

- Boolean lattices (YELLOW_2 and YELLOW_5),

---

[5]The *way below* relation is the key concept in continuous lattices - it is used to characterize continuous lattices.

- duality in lattices (YELLOW_7),

- modular and distributive lattices (YELLOW11),

- Moore-Smith convergence (YELLOW_6),

- Baire spaces and sober spaces (YELLOW_8),

- bases of topologies (YELLOW_9, YELLOW13, and YELLOW15),

- refinements of topologies (YELLOW_9),

- Hausdorff spaces (YELLOW12),

- product of topological spaces (YELLOW14),

- topological and poset retracts (YELLOW16).

The formalization was a stress test for the MIZAR software. It detected some errors and forced adjusting a number of quantitative parameters. The formalization would not be possible without cooperation with system designers and the Library Committee in improvement of software and a number of revisions to the MML.

4   Defining Lattices in MIZAR

There are at least two approaches to lattices in mathematics. According to the first, a lattice is an algebra with two binary operations ⊔ and ⊓ which satisfy the conditions of idempotency, associativity, commutativity, and absorption. According to the second, a lattice is a partially ordered set (poset) with suprema and infima for non empty finite subsets. Both approaches were already present in the MML and the correspondence between them was proved [19, 16, 1]. The second approach gives wider usage and is easier to generalize (e.g. by weakening the condition of partial ordering). This approach was chosen and the first revision of the MML consisted in generalization of posets and some lattice-theoretical concepts.

RelStr is the base structure of quasi ordered sets, posets, semilattices, and lattices and was introduced in [16] as follows:

```
definition
  struct (1-sorted) RelStr (#
    carrier     -> set,
    InternalRel -> Relation of the carrier
  #);
end;
```

If R is RelStr then R is a structure with at least 2 fields: carrier and InternatRel. A structure S can be a RelStr and may have more fields when its type is derived from RelStr. This is not the case when S is strict RelStr. The definition of the attribute strict is generated automatically by each structure definition.

The concept of a poset was introduced as follows:

```
definition
    mode Poset is reflexive transitive antis ymmetric RelStr;
end;
```

The attributes reflexive, transitive, and antisymmetric and the following existential cluster registration were introduced earlier.

```
definition
  cluster non em pty reflexive transitive antis ymmetric strict RelStr;
  existence
    proof
     :: Demonstration that such an ob ject exists
    end;
end;
```

(Two colons :: start a comment which ends at the end of the line.)

The above cluster assures existence of a RelStr type objects with any subset of the listed attributes.

For convenience and to be closer to usual notation the following definition was introduced.

```
definition
    let R be RelStr;
    let x, y be Element of the carrier of R;
    pred x <= y means
 :: ORDERS_1:def 9
     [x,y] in the InternalRel of R;
    synonym y >= x;
end;
```

The characterizations of reflexivity, transitivity, and antisymmetry were given in [2] as redefinitions:

```
definition
    let A be non em pty RelStr;
  redefine
    attr A is reflexive means
 :: YELLOW_0:def 1
     for x being Element of A holds x <= x;
```

```
    compatibility proof .... end;
end;

definition
  let A be RelStr;
 redefine
  attr A is transitive means
:: YELLOW_0:def 2
    for x,y,z being Element of A st x <= y & y <= z holds x <= z;
    compatibility proof .... end;
  attr A is antisymmetric means
:: YELLOW_0:def 3
    for x,y being Element of A st x <= y & y <= x holds x = y;
    compatibility proof .... end;
end;
```

The concept of lattice was introduced by definitions:

```
definition
  let R be RelStr;
  attr R is with_join means
:: LATTICE3:def 10
    for x,y being Element of R
      ex z being Element of R st x <= z & y <= z &
       for z' being Element of R st x <= z' & y <= z' holds z <= z';
  attr R is with_meet means
:: LATTICE3:def 11
    for x,y being Element of R
      ex z being Element of R st z <= x & z <= y &
       for z' being Element of R st z' <= x & z' <= y holds z' <= z;
end;

:: WAYBEL_0
definition
 mode Semilattice is with_meet Poset;
 mode sup-Semilattice is with_join Poset;
 mode LATTICE is with_join with_meet Poset;
end;
```

5   Continuous Lattices

The concept of directed sets was changed in MIZAR formalization. A directed set is non empty in mathlore and in the compendium. However, it happens often that we need a set which is directed or empty. MIZAR does not allow to write a type as *directed or empty set* and we decided to formalize the concept as follows:

```
definition
   let L be RelStr;
   let X be Subset of L;
   attr X is directed means
:: WAYBEL_0:def 1       :: CCL, Definition 1.1,  p. 2
     for x,y being Element of L st x in X & y in X
        ex z being Element of L st z in X & x <= z &  y <= z;
   attr X is filtered means
:: WAYBEL_0:def 2        :: CCL, Definition 1.1,  p. 2
     for x,y being Element of L st x in X &  y in X
        ex z being Element of L st z in X & z <= x & z <=  y;
end;
```

The theorem explaining correspondence to usual meaning has been proved also.

```
theorem :: WAYBEL_0:1
 for L being non empty transitive RelStr, X bein g Subset of L holds
   X is non empty directed iff
     for Y being finite Subset of X
       ex x being Element of L st x in X & x is_>=_than Y
   proof .... end;
```

The concept of completeness presented in [7] depends on a context. A complete poset, complete semilattice, and complete lattice satisfy different conditions. In MIZAR we introduced attributes

- `up-complete` as completeness with respect to directed sups,

- `inf-complete` as completeness with respect to non empty infs,

- `complete` as completeness with respect to all sups.

Then, in MIZAR notation

Compendium	MML
a *complete poset*	`up-complete Poset`
a *complete semilattice*	`inf-complete up-complete Semilattice`
a *complete lattice*	`complete LATTICE.`

The fact that a complete lattice is a complete poset and a complete semilattice is expressed in MIZAR (see [3]) by conditional cluster registration:

```
definition
  cluster complete -> up-complete inf-complete
                              (non empty reflexive RelStr);
    coherence
     proof
       let R be non empty reflexive RelStr;
       assume R is complete;
       ....
       thus R is up-complete by ...
       ....
       thus R is inf-complete by ...
     end;
 end;
```

The conditional registration is used automatically by MIZAR. Attributes up-complete and inf-complete are added to a type when it widens to non empty reflexive RelStr and already includes attribute complete. The concept of continuous lattices presented in the compendium depends on context. We decided to formalize it in as general way as possible because all meanings of it may be expressed by the basic continuous attribute and some extra conditions of completeness.

```
definition
  let L be non empty reflexive RelStr;
  attr L is continuous means
:: WAYBEL_3:def 6
  (for x being Element of L holds waybelow x is non empty directed) &
  L is up-complete satisfying_axiom_of_approximation;
end;
```

The attribute satisfying_axiom_of_approximation is introduced as follows.

```
definition
  let L be non empty reflexive RelStr;
  attr L is satisfying_axiom_of_approximation means
:: WAYBEL_3:def 5
    for x being Element of L holds x = sup waybelow x;
end;
```

The sup is the supremum operation, see [2]. The waybelow x is a set of all elements of L which are way below x, see [4].

The MIZAR notation for continuous posets:

Compendium	MML
a *continuous poset*	continuous up-complete Poset
a *continuous semilattice*	continuous up-complete Semilattice
a *complete-continuous semilattice*	continuous inf-complete up-complete Semilattice
a *continuous lattice*	continuous complete lattice

As the test of the correctness of the introduced concepts, the correspondence between locally compact topological spaces and continuous lattices has been proved. This correspondence is expressed by two theorems:

```
theorem :: WAYBEL_3:42
 for T being non empty TopSpace
     st T is_T3 & InclPoset(the topology of T) is continuous
  holds T is locally-compact;
```

```
theorem :: WAYBEL_3:43
 for T being non empty TopSpace st T is locally-compact
  holds InclPoset(the topology of T) is continuous;
```

The `InclPoset(the topology of T)` is the poset of open sets from space T ordered by inclusion.

6 Mixing Order and Topology

Topologies on posets induced by the ordering and, conversely, partial orders on topological spaces generated by topology are investigated in the theory of continuous lattices. For example, *Scott topology* introduced in the compendium is the family of sets which are inaccessible by directed sups. *Lawson topology* and *lower topology* are another example of such topologies. *Lawson topology* is the common refinement of *Scott* and *lower topologies* and *lower topology* is generated by complements of principal filters as subbasic open sets.

When investigating such topologies we need to use both theories: posets and topological spaces. In the case of *Lawson topology* we have in the same time three topologies and a poset. The solution from the compendium consists in introducing new notation like *Scott open*, *Scott closed*, *Scott neighbourhood*, etc. It is possible to do the same in MIZAR but such notation causes substantial technical difficulties with the use of general topology developed in the MML. Besides, such notation is not consequently applied in the compendium.

The problem was solved by multi prefixed structure definition in [11] and by mode definition in [5].

```
:: WAYBEL_9
struct(TopStruct, RelStr)
 TopRelStr (# carrier -> set,
               InternalRel -> (Relation of the carrier),
               topology -> Subset-Family of the carrier #);

definition
 let R be RelStr;
 mode TopAugmentation of R -> TopRelStr means
:: YELLOW_9:def 4
  the RelStr of it = the RelStr of R;
  existence proof .... end;
end;
```

TopStruct has two fields: `carrier` and `topology` and is the base structure of topological spaces. The structure `TopRelStr` is both the structure `TopStruct` and the structure `RelStr`. We may apply to it attributes defined for posets and attributes defined for topological spaces as well. If X is `TopRelStr`, then the `RelStr` of X will be `strict RelStr` and, moreover,

```
the RelStr of X = RelStr(# the carrier of X, the InternalRel of X #)
```

(analogically, for `TopStruct`).

```
:: WAYBEL_9
definition
 mode TopLattice is with_join with_meet reflexive transitive
         antisymmetric TopSpace-like TopRelStr;
end;
```

As an illustration of applied convention, let us compare the proposition 1.6 from the compendium, page 144, and corresponding MIZAR theorems.

    1.6. PROPOSITION. *Let* L *be a complete lattice.*
    (i) *An upper set* U *is Lawson open iff it is Scott open;*
    (ii) *A lower set is Lawson-closed iff it is closed under sups of directed sets.*

```
theorem :: WAYBEL19:41
:: 1.6. PROPOSITION (i), p. 144
 for S being Scott complete TopLattice
 for T being Lawson correct TopAugmentation of S
 for A being upper Subset of T st A is open
 for C being Subset of S st C = A holds C is open;

theorem :: WAYBEL19:42
```

```
:: 1.6. PROPOSITION (ii), p. 144
 for T being Lawson (complete TopLattice)
 for A being lower Subset of T holds
   A is closed iff A is closed_under_directed_su ps;
```

The implication from right to left in point (i) is proved in more general case:

```
theorem :: WAYBEL19:37
 for S being Scott complete TopLattice
 for T being Lawson correct To pAugmentation of S
 for A being Subset of S st A is o pen
 for C being Subset of T st C = A holds C is o pen;
```

7   Some Statistics

The project started in 1996. The compendium contains 334 pages and the theory formalized by the end of February 2000 covers about 180 pages of it (about 54% without taking into account the articles currently under development).

The following summarizes the number of articles from this project submitted to MML:

year	1996	1997	1998	1999	2000	1996–2000
YELLOW	8	1	5	3	0	17
WAYBEL	10	6	8	4	4	32
All	18	7	13	7	4	49
Y÷All %	44%	14%	38%	42%	0%	35%

The last line gives percentage of YELLOW series. This percentage is less than we expected.

	MML	WAYBEL	YELLOW	W&Y	Percentage
Articles	633	32	17	49	7.74%
Theorems	29,514	1,391	834	2225	7.54%
ave per art	46.6	43.5	49.1	45.4	–
Definitions	5,389	246	105	351	6.51%
ave per art	8.5	7.7	6.2	7.2	–
Size (kB)	46,966	2,867	1,273	4,140	8.82%
ave per art	74.2	89.6	74.9	84.5	–

The last column gives percentage of this project in the entire MML. Average numbers of theorems, definitions, and kilobytes show that the project is close to the MML average. Note the smaller average number of definitions which may indicate that the theory is explored more intensively.

The interaction between the project and the rest of the MML may be measured by the number of references between them. Each reference to the theorem coming from another article is called an external reference.

External references	Count	Percentage
All	317427	100.00%
All to Y&W	11677	3.68%
Outside of Y&W to Y&W	349	0.11%
In Y&W	26747	100.00%
In Y&W to Y&W	11328	42.35%

57.65% of all external references from the YELLOW and WAYBEL articles is to the rest of the MML. This indicates that the MML contained a substantial quantity of definitions and facts needed for our project.

There is, unfortunatly, no statistics concerning the quantity of work needed to formalize this material. However, we may state that it vary on authors and WAYBEL series needed much more work per line than YELLOW series.

8   Conclusions

Our main conclusion is that the MIZAR system seems satisfactory to formalize advanced mathematics.

The second conclusion is that the MML was satisfactorily rich to start formalization of the compendium. The YELLOW series constitutes only 35% of the whole project.

Formalization in MIZAR is still not as simple as doing mathematics traditionally. It should be improved in near future. Now, however, there are some gains. The results are mechanically checked. There is an automatic access to the knowledge stored and the net of concepts is explicit. (This helped very much for new authors to start.) The information may be mechanically explored: changed, generalized, and edited. Reorganization of a machine readable mathematical text is much easier than reorganization of such a text written on paper. (Such reorganizations were quite often required in our project.)

The work done in this project resulted in numerous improvements of the MIZAR system and, also, it revealed a number of issues that are investigated:

- tools for searching MML: semantic searching which can distinguish homonyms, glue synonyms, and recognize hidden arguments,

- proof assistance based on the exploration of existing proofs in the MML and computer algebra,

- reorganization of the MML: revisions of existing articles and "online revision" mechanism available by environment directives,

- development of MIZAR language to improve the convenience of formulation of definitions and proofs, the length of proofs, and the flexibility of type structure: type modifier, attributes with explicit arguments, new realization of structure types.

References

[1] Grzegorz Bancerek. Complete lattices. *Formalized Mathematics*, 2(**5**):719–725, 1991. MML: LATTICE3.

[2] Grzegorz Bancerek. Bounds in posets and relational substructures. *Formalized Mathematics*, 6(**1**):81–91, 1997. MML: YELLOW_0.

[3] Grzegorz Bancerek. Directed sets, nets, ideals, filters, and maps. *Formalized Mathematics*, 6(**1**):93–107, 1997. MML: WAYBEL_0.

[4] Grzegorz Bancerek. The "way-below" relation. *Formalized Mathematics*, 6(**1**):169–176, 1997. MML: WAYBEL_3.

[5] Grzegorz Bancerek. Bases and refinements of topologies. *Formalized Mathematics*, 7(**1**):35–43, 1998. MML: YELLOW_9.

[6] Ewa Bonarska. *An Introduction to PC Mizar*. Fondation Philippe le Hodey, Brussels, 1990. Revised version, 2000: http://merak.uwb.edu.pl/~bancerek/introduction/.

[7] G. Gierz, K. H. Hofmann, K. Keimel, J. D. Lawson, M. W. Mislove, and D. S. Scott. *A Compendium of Continuous Lattices*. Springer-Verlag, Gerlin, Heidelberg, 1980.

[8] Andrzej Grzegorczyk. *Zarys arytmetyki teoretycznej*. Biblioteka Matematyczna. PWN, Warszawa, 1971.

[9] Stanisław Jaśkowski. *On the Rules of Supposition in Formal Logic*. Studia Logica. Warsaw University, 1934. Reprinted in S. McCall, Polish Logic in 1920–1939, Clarendon Press, Oxford.

[10] Peter T. Johnstone. *Stone Spaces*. Cambridge University Press, Cambridge, London, New York, 1982.

[11] Artur Korniłowicz. On the topological properties of meet-continuous lattices. *Formalized Mathematics*, 6(**2**):269–277, 1997. MML: WAYBEL_9.

[12] E. G. H. Landau. *Grundlagen der Analysis*. Akademische Verlag, Leipzig, 1930.

[13] Piotr Rudnicki and Andrzej Trybulec.   On equivalents of well-foundedness. *Journal of Automated Reasoning*, 23(3-4):197–234, 1999.

[14] Andrzej Trybulec. Built-in concepts. *Formalized Mathematics*, 1(**1**): 13–15, 1990. MML: AXIOMS.

[15] Andrzej Trybulec. Tarski Grothendieck set theory. *Formalized Mathematics*, 1(**1**):9–11, 1990. MML: TARSKI.

[16] Wojciech A. Trybulec. Partially ordered sets. *Formalized Mathematics*, 1(**2**):313–319, 1990. MML: ORDERS_1.

[17] L. S. van Benthem Jutting. Checking Landau's "Grundlagen" in the Automath system, 1977. PhD thesis.

[18] Freek Wiedijk. *Mizar: An Impression*, 1999. http://www.cs.kun.nl/~freek/mizar/.

[19] Stanisław Żukowski. Introduction to lattice theory. *Formalized Mathematics*, 1(**1**):215–222, 1990. MML: LATTICES.

Ω-ANTS – An Open Approach at Combining Interactive and Automated Theorem Proving

Christoph Benzmüller*   Volker Sorge†

Abstract. *We present the Ω-ANTS theorem prover that is built on top of an agent-based command suggestion mechanism. The theorem prover inherits beneficial properties from the underlying suggestion mechanism such as run-time extendibility and resource adaptability. Moreover, it supports the distributed integration of external reasoning systems. We also discuss how the implementation and modeling of a calculus in our agent-based approach can be investigated wrt. the inheritance of properties such as completeness and soundness.*

1 Introduction

We present the new Ω-ANTS automated theorem proving approach that is build on top of the Ω-ANTS agent-based command suggestion mechanism. This mechanism has been originally developed to support the user in interactive theorem proving by using available resources in-between user interactions to search for the next possible proof steps [4]. This is done via a hierarchical blackboard-architecture where agents concurrently check for applicable commands (i.e. commands that apply proof rules) and the most promising commands are dynamically presented to the user. The faster a command's applicability can be analysed the faster it will be reported immediately to the user. Further benefits of the distributed Ω-ANTS command suggestion mechanism are the increased robustness (errors in the distributed computations do not harm the overall mechanism), its resource- and user-adaptability, and its run-time extendibility and modifiability [5].

---
*The author would like to thank EPSRC for its support by grant GR/M99644.
†The author's work was supported by the 'Studienstiftung des deutschen Volkes'.

In this paper we present how we achieve the automation of Ω-ANTS. On the one hand the concurrency enables the integration of external reasoning systems into the suggestion process. External reasoners can either be used to suggest whole subproofs or to compute particular arguments of commands. On the other hand Ω-ANTS can be automated directly by executing suggested commands automatically instead of just presenting them to the user. Thereby it is important to restrict the set of involved commands to those suitable for automation and to fix a certain clock speed determining the period of time the suggestion mechanism may maximally consume for its computations in-between the automated command executions. In any proof state where Ω-ANTS cannot find any new applicable commands it simply backtracks by retracting the lastly executed command.

The Ω-ANTS suggestion mechanism and the Ω-ANTS theorem prover have been developed and implemented within the ΩMEGA theorem proving environment [17]. However, the approach is not restricted to a particular logic, calculus, or theorem proving environment. It can be rather seen as an approach parameterised over the particular calculus it is working for. In this respect the question arises how the designer of the Ω-ANTS agents which have to be provided for each calculus rule can ensure that the modeling guarantees a complete proof search in Ω-ANTS. This question is discussed in the second half of the paper by informally defining properties of agent societies in Ω-ANTS which are necessary to ensure completeness and by giving some examples how these properties are checked in practice.

This paper is organised as follows: Section 2 sketches the Ω-ANTS command suggestion mechanism (for further details see [4, 5]), illustrates its declarative agent specification language, and sketches a formal semantics. Section 3 describes how external reasoners can be integrated at different layers. In Section 4 the Ω-ANTS theorem prover built on top of the suggestion mechanism is introduced and completeness aspects are discussed in Section 5. We conclude with discussing some related work and hinting at future work.

2 The Ω-ANTS Suggestion Mechanism

In this section we sketch the hierarchical, agent-based suggestion mechanism underlying the Ω-ANTS theorem prover. We also discuss the declarative agent specification language supporting the specification and modification of agents at run-time, and sketch how this language can be linked to a formal semantics.

Agent-based architecture The suggestion mechanism originally aims at supporting a user in interactive tactical theorem proving to choose an appropriate proof rule from the generally large set of available ones. It computes and proposes commands that invoke proof rules that are applicable in a given proof state.[1] This is basically done in two steps: firstly, by computing whether there are any possible instantiations for single arguments of a command in the current proof state; and secondly, by gathering those commands for which at least some arguments could be instantiated and presenting them in some heuristically ordered fashion to the user.

An important notion for the Ω-ANTS mechanism is that of a *Partial Argument Instantiations (PAI)* for a command. Considering a command and its corresponding proof rule there is usually a strong connection between the formal arguments of both, i.e. the formal arguments of the command are generally a subset of the formal arguments of the proof rule. As an example we observe the proof rule $\wedge$I and its corresponding command AndI:

$$\frac{A \quad B}{A \wedge B} \wedge \text{I} \longrightarrow \frac{\text{LConj} \quad \text{RConj}}{\text{Conj}} \text{ AndI}$$

Here the command's formal argument Conj needs to be instantiated with an open proof node containing a conjunction, LConj and RConj with nodes containing the appropriate left and right conjuncts, respectively. In general, a command's formal arguments need to be partially instantiated only, in order to be applicable. For instance, AndI is also applicable if only the Conj argument is provided, resulting in the introduction of two new open proof nodes containing the two conjuncts. Or additionally one of LConj or RConj or even both could be provided, resulting in the introduction of only one open node or in simply closing the given open conjunction. Thus, we can denote partial argument instantiations for a command as a set relating some of a command's formal argument to actual arguments for its execution. One possible PAI for AndI would be $(\text{Conj} : x, \text{LConj} : y)$ where x and y are proof nodes that contain the appropriate formulas. PAIs can also be seen as functions, indexed by the different command names, with the set of argument-names as domain and the infinite set of possible proof lines and parameters as codomain. For instance, PAIs for AndI can be represented as particular functions

$$\text{PAI}^{\text{AndI}} : \{\text{Conj, LConj, RConj}\} \longrightarrow \text{Prooflines} \cup \text{Parameters} \cup \{\epsilon\}$$

where ϵ is a special symbol denoting the empty proofline. In these sense the PAI $(\text{Conj} : x, \text{LConj} : y)$ for AndI is realised by a respective function such that $\text{PAI}^{\text{AndI}}(\text{Conj}) = x$, $\text{PAI}^{\text{AndI}}(\text{LConj}) = y$, and $\text{PAI}^{\text{AndI}}(\text{RConj}) = \epsilon$.

---

[1]For the remainder of the paper, if we talk about applicability of a command we always mean the applicability to the corresponding proof rule (e.g., a calculus rule, a tactic, a proof planning method, or an external system call) in the given proof state.

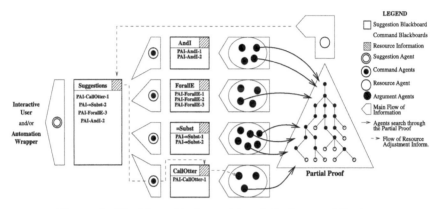

Figure 1. The hierarchical, agent-based Ω- ANTS-architecture.

The idea of the suggestion mechanism is to compute in each proof state for each command PAIs as complete as possible, to determine which commands are applicable, and then to give preference to those, e.g., with the most complete PAIs. The first task is done by societies of *Argument Agents* (rightmost circles in Figure 1) where one society is associated with each of the commands. Each argument agent is associated with one or several of the command's formal arguments and has a specification for possible instantiations of these arguments. Its task is to search for proof nodes in the partial proof or to compute parameters according to its specification. Argument agents exchange results via *Command Blackboards* (for each command one command blackboard is provided) using PAIs as messages. Every argument agent commences its computations only when it finds a PAI on the command blackboard that contains instantiations of arguments that are relevant for its own computations.

For example, the AndI argument agent associated with Conj searches the partial proof for an open node containing a conjunction and, once it has found one, say in node x, it places a respective PAI (Conj : x) on the command blackboard. Now the agents for LConj and RConj can use this result in order to look simultaneously for nodes in the given partial proof containing the appropriate left or right conjunct. Each argument agent only reads old suggestions and possibly adds expanded new suggestions, thus there is no need for conflict resolutions between the agents.

On top of the layer of argument agents are the *Command Agents* (dotted circles). Their task is to monitor the command blackboard associated with the command and to heuristically order the PAIs from most promising (e.g., most complete) to least promising. Whenever their heuristics indicate that there is a new best PAI on the command blackboard they pass it to the

Suggestion Blackboard. The suggestion blackboard itself is again monitored by the *Suggestion Agent* (leftmost double circle) which sorts the entries with respect to its heuristic criteria and presents them to the user.

When the Ω-Ants mechanism is started all command blackboards are initialised with the empty PAI. The agents then autonomously search for applicable commands and the newest suggestions are successively presented to the user. At any point a command can be executed and when the proof state has actually been changed the Ω-Ants mechanism is re-initialised in order to compute new suggestions for the modified proof state. Ω-Ants can also be used to respond to particular user queries, i.e. the user can interactively specify certain argument instantiations and the mechanism tries to complete these.

The whole mechanism can be adjusted during run-time by changing sorting heuristics for the command blackboards and the suggestion blackboard or by removing, adding, or modifying argument agents. Moreover, Ω-Ants employs a resource mechanism that automatically disables and enables argument agents with respect to their usefulness and performance in particular proof states. Although not depicted here, the mechanism also contains classification agents whose purpose is to classify the focused subproblem in terms of logic and mathematical theory is belongs to. This information is communicated within the blackboard architecture enabling agents to decide whether they are appropriate (i.e. should be active) in the current proof state or not. See [5] for further details.

A Declarative Agent Specification Language   In Ω-Ants only the argument agents need to be explicitly specified. All other agents are then generated automatically (certain heuristics may be adapted by the user, though). Argument agents are implemented with a Lisp-like declarative language such as the following two argument agents for the **AndI** command:

```
𝔄₁:
(agent~defagent AndI c-predicate
   (for Conj) (uses )
   (exclude LConj RConj)
   (definition
     (logic~conjunction-p Conj)))
```

```
𝔄₄:
(agent~defagent AndI s-predicate
   (for RConj) (uses Conj)
   (definition
     (logic~right-conjunct-p RConj Conj))))
```

The agent $\mathfrak{A}_1$ is defined as a **c-predicate** agent, indicating that it will always restrict its search to open proof nodes, i.e., possible conclusions. **s-predicate** agents like $\mathfrak{A}_4$ in contrast search the support nodes for possible premises. The proof nodes $\mathfrak{A}_1$ is looking for are instantiations of the argument **Conj**, given in the **for**-slot. The empty **uses**-slot indicates that the agent does not require any already given argument suggestions in a PAI for its computations. The **exclude**-slot on the other hand determines that this agent must not complete any PAI that already contains an in-

$\mathfrak{A}_1: \quad \mathfrak{C}^{\{Conj\}}_{\{\},\{LConj,RConj\}} \quad := \quad \lambda Conj_{\bullet}(Conj \equiv A \wedge B)$

$\mathfrak{A}_2: \quad \mathfrak{C}^{\{Conj\}}_{\{LConj\},\{RConj\}} \quad := \quad \lambda Conj_{\bullet}(Conj \equiv A \wedge B) \,\&\, (LConj \equiv A)$

$\mathfrak{A}_3: \quad \mathfrak{C}^{\{Conj\}}_{\{RConj\},\{LConj\}} \quad := \quad \lambda Conj_{\bullet}(Conj \equiv A \wedge B) \,\&\, (RConj \equiv B)$

$\mathfrak{A}_4: \quad \mathfrak{S}^{\{RConj\}}_{\{Conj\},\{\}} \quad := \quad \lambda RConj_{\bullet}(Conj \equiv A \wedge B) \,\&\, (RConj \equiv B)$

$\mathfrak{A}_5: \quad \mathfrak{S}^{\{LConj\}}_{\{Conj\},\{\}} \quad := \quad \lambda LConj_{\bullet}(Conj \equiv A \wedge B) \,\&\, (LConj \equiv A)$

$\mathfrak{A}_6: \quad \mathfrak{C}^{\{Conj\}}_{\{LConj,RConj\},\{\}} \quad := \quad \lambda Conj_{\bullet}(Conj \equiv A \wedge B) \,\&\, (LConj \equiv A) \,\&\, (RConj \equiv B)$

Figure 2. A society of argument agents for command `AndI`.

stantiation for arguments `LConj` or `RConj`. In the special case of $\mathfrak{A}_1$ this means the agent is exactly triggered by the empty PAI. The idea for this exclusion constraint is to suppress redundant or even false computations.

The full set of argument agents for the `AndI` command is given in Figure 2 in a specification meta-language. `c-predicate` and `s-predicate` agents are denoted by $\mathfrak{C}$ and $\mathfrak{S}$ respectively, the superscript set corresponds to the `for`-list, and the `uses`- and `exclude`-list to the first and second index. The subset of the nodes in a partial proof that will be detected by each argument agent can be formally described by a λ-term (characteristic function). When running over the partial proof the agents use these characteristic functions to test each node before possibly returning an expanded PAI. A and B are free meta-variables. $\equiv$ and $\&$ are symbols of the meta-language with the meaning, for instance in agent $\mathfrak{A}_2$, that given an arbitrary formula A instantiating argument `LConj` then $Conj$ has to be of form $A \wedge B$, i.e. the left hand side of $Conj$ is determined by the already given suggestion `LConj` whereas is right hand side is still *free*.

This attempt at a formal semantics for our agent definitions by assigning characteristic functions to them does not yet address the agents functional behaviour (they pick up & return potentially modified PAIs) nor does it formally regard the uses and exclude-restrictions. This is the idea of the λ-expression for agent $\mathfrak{A}_2$ below. Assuming that PAIs are represented as functions this term denotes that $\mathfrak{A}_2$ picks up certain PAIs on the blackboard and returns possibly modified ones while using an (extended/modified) characteristic function in the previous sense as filter. Here the [.]-brackets denote a function which accesses the formula content of the proofline given as an argument to it (note that PAIs map argument names to prooflines, while here we want to talk about the formulas of the prooflines[2]).

$\lambda\, PAI \,_{\bullet}\, \lambda\, Conj_{\mathrm{open}}\, _{\bullet}$
     if $PAI(\mathrm{Conj}) \equiv \epsilon \,\&\, PAI(\mathrm{LConj}) \not\equiv \epsilon \,\&\, PAI(\mathrm{RConj}) \equiv \epsilon$

---

[2]In other parts of this paper we do not take this so seriously and assume that the user recognises whether we address a proof line or its formula content from the context.

then   if $[Conj] \equiv$ A $\wedge$ B & $[PAI(\text{LConj})] \equiv$ A
       then $PAI|_{\{\text{LConj, RConj}\}} \cup \{\text{Conj} \mapsto Conj\}$ $\rightarrow$ new ext. PAI
       else PAI $\rightarrow$ no new PAI
       fi
else PAI $\rightarrow$ no new PAI
fi

3 Integration of External Reasoning Systems

The following four examples illustrate how external reasoners can be integrated into Ω-ANTS. The first row presents four inference rules and the second the corresponding commands which we want to model in Ω-ANTS.

$$\frac{P_1 \quad \cdots \quad P_n}{C} \; Otter \qquad \frac{A \quad B \Rightarrow C}{C} \quad \begin{matrix} mp\text{-}mod\text{-}Otter(A \Rightarrow B) \\ mp\text{-}mod\text{-}CAS(A \overset{\text{simpl}}{\rightarrow} B) \end{matrix}$$
                  Mace

$$\frac{\text{Prem}_1 \quad \cdots \quad \text{Prem}_n}{\text{Conc}} \; \begin{matrix} Otter \\ Mace \end{matrix} \qquad \frac{\text{Left} \quad \text{Impl}}{\text{Conc}} \quad \begin{matrix} \text{mp-mod-Otter(Impl-Prob)} \\ \text{mp-mod-CAS(Simpl-Prob)} \end{matrix}$$

The first two rules describe the integration of the first-order theorem prover OTTER and the propositional logic decision procedure MACE. These commands may be used in a given proof state in order to justify a goal from its premises by the application of one of these external systems. The next two rules describe a situation where external reasoners are used within an inference modulo, in our particular case modus ponens modulo the validity of an implication (to be checked by OTTER) and modus ponens modulo the simplifiability of a proposition (to be analyzed by a computer algebra system). For instance, sensible instances of these commands in a concrete proof situation would be: Left $\leftarrow \forall x.p(x) \wedge q(x)$, Impl $\leftarrow p(a) \Rightarrow r(a)$, Conc $\leftarrow r(a)$, and Impl-Prob $\leftarrow (\forall x.p(x) \wedge q(x)) \Rightarrow p(a)$ for mp-mod-Otter, and Left $\leftarrow$ continuous$(\lambda x.1 - \cos^2(x))$, Impl $\leftarrow$ continuous$(\lambda x.\sin^2(x)) \Rightarrow$ something$(\lambda x.\sin^2(x))$, Conc $\leftarrow$ something$(\lambda x.\sin^2(x))$, Simpl-Prob$\leftarrow$continuous $(\lambda x.1 - \cos^2(x)) \overset{\text{simpl}}{\rightarrow}$ continuous$(\lambda x.\sin^2(x))$ for mp-mod-CAS. The idea is that the external systems are used to check the 'modulo'-side-conditions of these rules. Note, that in contrast to the theorem proving modulo approach described in [9] we explicitly facilitate and support the integration of non-decision procedures; a strict separation of deduction and computation is not needed due to the distribution and resource-guidance aspect of the Ω-ANTS mechanism.

We are here not concerned with correctness issues for the integration of the external systems. However, since we are working in the ΩMEGA environment we can make use of the work already done in this area that ensures the correctness by translating proofs or computations from external reasoners into primitive inference steps of ΩMEGA [16, 19].

If we, for example, consider the `Otter` and the `mp-mod-CAS` command we can observe two different ways of integrating the external reasoners into agents: For the `Otter` command one agent attacks the focused open goal in-between user interactions and as soon as OTTER finds a proof the application of this command is suggested to the user. Thus, the agent employs the external system to prove an open sub-problem. Similarly, other external reasoners can be integrated.

In case of the `mp-mod-CAS` command an agent will first look for appropriate implication proof nodes with respect to the open goal. This agent's results trigger another agent that which employs an integrated computer algebra system to look for appropriate proof nodes as instances for argument `Left`. More precisely, the latter agent checks whether a proof node can be matched with the antecedent of the implication with respect to algebraic simplification of sub-terms. Hence, the agent uses an external reasoner only to find possible instantiations of arguments.

4 Automation

The Ω-ANTS suggestion mechanism of Section 2 can be automated into a full-fledged proof search procedure by embedding the execution of suggested commands into a backtracking wrapper. The algorithm is given in Figure 3.

The basic automation performing a depth first search is straight forward: The suggestion mechanism waits until all agents have performed all possible computations and no further suggestions will be produced and then executes the heuristically preferred suggestion (1a&2). When a proof step is executed and the proof is not yet finished, the remaining suggestions on each command blackboard are pushed on the backtracking stack (3). In case no best suggestion could be computed Ω-ANTS backtracks by popping the first element of the backtrack stack and re-instantiating its values on the blackboards (6). The proof is constructed as an explicit proof plan data structure of ΩMEGA [8]. It enables to store proofs in a generalised natural deduction format, i.e. proof steps cannot only be justified by basic natural deduction rules but by abstract tactics or computations of external reasoners as well. Moreover, the proof plan data structure supports the expansion of externally computed proofs into primitive inference steps and thus the check for correctness as well as the storage of information for the automation loop directly in the proof object.

The simple automation loop is complicated by the distinct features of Ω-ANTS: (i) some agents can perform infinite or very costly computations, (ii) commands can be executed by the user parallel to the automation, and (iii) the components of Ω-ANTS can be changed at run-time. Further-

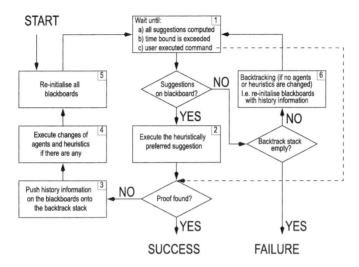

Figure 3. Main-loop of the Ω- ANTS theorem prover.

more, the automation can be suspended and revoked especially in order to perform the latter two interaction possibilities in a coordinated way.

We avoid that Ω-ANTS is paralysed by agents that get stuck in infinite computations by giving a time limit after which the best command, suggested so far, is executed (cf. step 1b). However, such a proof step is treated special when backtracking, since then the blackboards will be re-instantiated with all the values of the proof step, i.e. containing the executed command as well. This way there is a second chance for agents that could not contribute the first time to add information. The question how the Ω-ANTS theorem prover can avoid to get lost on an infinite branch in the search space without ever backtracking will be addressed in the completeness discussion in Section 5.

If a command has been executed by the user the loop proceeds immediately with saving the blackboards' history without executing another command (1c). When backtracking the whole history on the last step is re-instantiated onto the blackboards, possibly containing also the command executed by the user, in order not to loose possible proofs (1c&6).

One main feature of Ω-ANTS is its run-time adaptability by adding or deleting agents or changing the filter and sorting heuristics used by the suggestion and commands agents. These changes also take effect when running the automation wrapper (4). The automation wrapper can be suspended by the user at any time, for instance, in order to analyzed the current proof state and to add, change or remove certain agents from the suggestion mechanism. It can then be resumed using all the information computed so far.

We briefly summarise the user interaction facilities inherited by the Ω-ANTS prover from the Ω-ANTS suggestion mechanism:

Pure user interaction/mixed initiative reasoning: In automation mode the entries on the suggestion blackboard are (theoretically[3]) steadily visible to the user, who can interfere with the automation wrapper by executing a command before the automation wrapper does.

Adjustment of resource bounds: The user may want to actively modify the resource bounds (time, memory, deactivation threshold) in order to adapt the system to particular needs.

Disable/resume single agents: Ω-ANTS allows to disable/resume single agents, agent societies, or the whole mechanism at run-time.

Modification/addition of argument agents: The user may want to specify and load new agents at run-time or modify the definition of already given agents. This is supported by the declarative agent-specification language.

Modification of command/suggestion agents: In order to influence the provers search through the search space the user may want to choose different heuristics and sorting criteria for these agents.

The Ω-ANTS system has been applied to automate the propositional logic fragment of the normal form natural deduction calculus NIC [7], see [2] for more details. We currently experiment with the full first-order fragment of NIC. The integration of external reasoners has been tested with the propositional logic prover MACE, the first-order provers OTTER and SPASS, the higher-order prover TPS, and the computer algebra system MAPLE. The theorems we are working with are still all relatively simple and nothing any of the involved systems is not able to solve on its own. The computations involved are mainly to solve equations and compute derivatives.

5 Ω-ANTS and Completeness

In this section we introduce and discuss some notions that are necessary to characterise and guarantee completeness and soundness of a theorem prover based on Ω-ANTS with respect to the underlying calculus. The discussion is rather informal since we have yet to define completely formal syntax and semantics for our agent specification language. However, the following

---

[3]In our experiments with the NIC calculus, the theorem prover is unfortunately much faster than the graphical user interface to allow a synchronised displaying.

shall both give an intuition for the properties that need to be considered and contributes to a better understanding of Ω-ANTS.

Given a theoretically complete calculus, how can it be modeled in Ω-ANTS such that completeness is still assured in the mechanism? Note, that we do not address the theoretical completeness of the underlying calculus itself, in fact we do not even need to specify here what particular logic and calculus we are interested in. We rather aim to ensure that each calculus rule application that is theoretically possible in a given proof state can indeed be determined and suggested by the Ω-ANTS mechanism. In particular we will discuss two different notions of completeness in this sense, namely *interaction completeness* and *automation completeness*. This is due to twofold bias of the Ω-ANTS system as a suggestion mechanism and as an automated theorem prover. The authors admit that naming these properties also 'completeness' might be slightly misleading. However, automation (interaction) completeness of the agent societies involved taken together with the 'theoretical (logical) completeness' of a calculus implies that a complete proof search is actually supported by Ω-ANTS.

Theoretical completeness investigations typically assume non-limited resources like computation time and space. In our case the resources available to the Ω-ANTS-system in-between the command executions are crucial wrt. completeness as well. However, for the time being we neglect points possibly interfering with this assumption, in particular cases 1(b) or 1(c) of the prover's main-loop in Figure 3 and the existence of agents with calls to undecidable procedures such as the OTTER agent in Section 3.

Automation Completeness Automation completeness depends in the first place on the *suggestion completeness* of the argument agent societies associated with each rule: A society of suggestion agents working for a single command C is called *suggestion complete* wrt. a given calculus, if in any possible proof state all PAIs of a command necessary to ensure completeness of the calculus can be computed by the mechanism. Under the resource abstraction assumption from above suggestion completeness requires that each particular agent society consists of *sufficiently* many individual suggestion agents and that their particular definitions are *adequate* wrt. the structural dependencies and side-conditions of the respective calculus rule. *Adequacy* basically excludes wrong agent specifications, while *Sufficiency* refers to the ability of an agent society to cooperatively compute each applicable PAI in a given proof state.

We call a command agent *non-excluding* if it indeed always reports at least one selected entry from the associated command blackboard to the suggestion blackboard as soon as the former contains some applicable PAIs. And the suggestion agent is non-excluding if it always reports the complete set of entries on the command blackboard to the automation wrapper. This ensures that computed PAIs are actually propagated in the mechanism.

We additionally have to ensure that the proof search is organised in a *fair* way by ensuring that the execution of an applicable PAI suggested within a particular proof step cannot be infinitely long delayed. The fairness problem of Ω-ANTS is exactly the same as in other theorem proving approaches performing depth first search. In our experiments with the propositional logic fragment of the NIC this problem did not occur as the considered fragment defines a decision procedure. However to ensure that the prover does not get lost on infinite search path when working with the full first-order fragment of NIC we chose iterative deepening search.

Our mechanism can then be called automation complete wrt. to a given calculus C if (i) the agent societies specified are suggestion complete wrt. C, and (ii) the command agents for C and the suggestion agent are non-excluding, (iii) the search procedure is fair and (iv) the resource bounds and deactivation threshold are chosen sufficiently high, such that each agents computation terminates within these bounds.

We illustrate the notions of adequacy and sufficency in more detail with the example of the AndI agents. We claim that the agents $\mathfrak{A}_1 \ldots \mathfrak{A}_6$ of Figure 2 are both (a) adequate and (b) sufficient to apply AndI (whenever possible) in automated proof search.

(a) To show that all computable suggestions are indeed applicable we check that each agent produces an adequate predicate if all arguments of the uses slot are instantiated correctly. We observe this in the case, of agent $\mathfrak{A}_2$ when applying it to a PAI of the form (LConj:a). Here a is an arbitrary but fixed term. The resulting predicate is $Conj \equiv a \wedge B$ which permits all conjunctions with left conjunct a and is therefore adequate.

After checking adequacy of all single agents we have to ensure adequacy of cooperation between agents. That is, to show that no incorrect PAIs can be assembled by cooperation of agents with correct predicates. Here we are only concerned with agents whose for-, uses-, and exclude-list does not contain all possible arguments of the command, thus in our case agents $\mathfrak{A}_4$ and $\mathfrak{A}_5$. It can be easily seen that even if, for instance, $\mathfrak{A}_4$ is applied to a PAI already containing an instantiation for LConj, adding an appropriate instantiations for RConj will maintain the PAI's applicability, provided it was correct to begin with.

(b) To ensure sufficiency we have to show that each PAI of AndI necessary for automation can (cooperatively) be computed. In automatic mode the NIC calculus is intended for pure backward search and thus the possible PAIs are of the form[4] i) (Conj:$a \wedge b$), ii) (Conj:$a \wedge b$, LConj:a), iii)

---

[4]PAIs are essentially sets and thus the order of the particular entries is not important.

(Conj:$a \wedge b$, RConj:b), or iv) (Conj:$a \wedge b$, LConj:a, RConj:b), where a and b are arbitrary but fixed formulas occurring in a partial proof P. We representatively discuss case ii) and verify that each PAI of form S = (Conj:$a \wedge b$, LConj:a) that is applicable in P will actually be computed. As S is applicable, P must contain an open node containing $a \wedge b$ together with a support node containing a. Initially the command blackboard contains the empty PAI () to which only $\mathfrak{A}_1$ can be applied. Provided the underlying implementation, i.e. the function logic~conjunction-p is correct, $\mathfrak{A}_1$'s predicate suffices to compute (Conj:$a \wedge b$). This PAI in turn triggers the computations of $\mathfrak{A}_4$ and $\mathfrak{A}_5$ with the respective instantiated predicates $RConj \equiv b$ and $LConj \equiv a$. Since the latter is true on the support node containing a, $\mathfrak{A}_5$ returns the PAI in question.

When checking all other cases we can observe that for the automation mode (where pure backward reasoning is assumed) the agents $\mathfrak{A}_1$,$\mathfrak{A}_4$, and $\mathfrak{A}_5$ are already sufficient. And indeed the other three agents are needed to support user interaction, only. For instance, the user can apply Ω-ANTS to complete a particular PAI like (LConj:a) which will trigger the computations of agent $\mathfrak{A}_2$.

Interaction Completeness   Interaction completeness of a calculus implies that one never has to rely on another interaction mechanism besides Ω-ANTS in order to perform possible proof steps within a given calculus. Therefore, we have to show that all possible PAIs to apply a rule interactively can be computed. This is generally a stronger requirement than for automation completeness as can be easily observed with our AndI example. When automated the NIC calculus strictly performs backward search and only the PAIs (i)—(iv) given above are legitimate. However, when using the calculus interactively forward reasoning (i.e. a PAI of the form (LConj:a, RConj:b)) is a perfectly legal option. But it can be easily seen that this PAI cannot be computed with the given agent society and thus $\{\mathfrak{A}_1 \ldots \mathfrak{A}_6\}$ are not interaction complete.

When dealing with interaction completeness we have also to consider all possible initialisations of the command blackboards. While in automation mode the blackboards are always initialised with the empty PAI, the user can ask Ω-ANTS interactively to complete a particular PAI (such as (LConj:a)) which is then used as initial value on the blackboard. It is necessary to show sufficiency and adequacy for all possible initialisations.

Soundness   Should not the *soundness* aspect be addressed here as well? Our answer is no, as we presuppose that the underlying theorem proving environment takes care of a sound application of its own proof rules. Furthermore, in systems such as ΩMEGA soundness is always only guaranteed on the level of primitive inferences and not necessarily for all proof methods etc. involved. Thus, soundness requirements when computing suggestions for methods that do not necessarily lead to a correct proof would not make sense. Thus, instead of *logical soundness* we are rather interested in the notion of *applicability*. This notion relates the PAIs computed by Ω-ANTS to the particular side-conditions of the underlying proof rules (whether they are logically sound or not).

The effect of non-applicable PAIs suggested to the user or the automation wrapper might lead to failure when applying the respective command. In the current implementation such a failure will simply be ignored and the responsible PAI is discarded. However, too many non-executable suggestions might negatively influence the mechanisms user-acceptance and especially the performance of the automation wrapper.

6   Related Work

There exist several theorem proving environments where a mixture of interactive and partial automated proving is supported. In systems such as PVS [18] and HOL [13] special tactics are available that can be used to automatically solve certain problems. These tactics are essentially proof procedures build on top of the primitive inferences of the respective systems but do not directly construct a proof in terms of primitive inferences, although the automated parts can be, at least in the case of HOL, expanded. Moreover, there is no possibility for a user of the system to change the behaviour of the automation tactic during its application. In the TPS system [1] interaction and automation can also be interleaved and any automatic proving attempt can be interrupted, its behaviour changed and restarted by the user. The automation is achieved by using a mating search technique that is substantially different from the natural deduction calculus that is used for interactive proving. Finally, an approach to achieve automation in an interactive environment is to enable the use of external reasoners which is, for instance, one of the features of the ΩMEGA system [17]. However, without the Ω-ANTS part, application of rules, tactics and external reasoners cannot be automated.

As an environment that is especially designed to support the combination of interactive and automated theorem proving together with the use of already existing reasoning systems, is the Open Mechanised Reasoning System [12, 11] that has been extended to facilitate computer algebra systems [6]. While the concept of a reasoning structure to represent explicit proof states is similar to our concept of a proof object, external reasoners are connected as *plug-and-play* components which requires significant changes to their control components and therefore complicates the use of existing technology.

7   Conclusion

We presented the Ω-Ants theorem prover build on top of the agent-based Ω-Ants suggestion mechanism. This theorem prover inherits interesting features from the underlying suggestion mechanism and due to the distribution of computations down to a very fine-grained layer (e.g. reasoning about potential instances of single arguments of the considered inference rules) it especially supports the integration of external reasoning systems at various layers. We have illustrated that the Ω-Ants architecture especially supports deduction modulo computation/deduction performed by external reasoners. As the same suggestion mechanism that supports user-interaction is now also used as the main part of the automated theorem prover's inference machine the architecture also supports a close integration of interactive and automated theorem proving. This is underlined by the various interaction facilities the Ω-Ants prover already supports. The system can be seen as an open approach that is parameterised over the particular calculus it is working for (and note that it is only in a technical sense restricted to the ΩMEGA environment in which it has been developed). The calculus it is working for can even be modified/extended at run-time, making our system in the long-run also interesting for the integration of components aiming at learning new inference rules from past proof experience [15]. The learned rules could then be dynamically added to the running system.

Immediate further work is a more rigorous formalisation of the agent specification language as well as to formally model the connection between Ω-Ants and underlying calculi. Other future work is to analyse whether our system could benefit from a dynamic agent grouping approach as described in [10] and whether it can fruitfully support the integration of proof critics as discussed in [14]. The Ω-Ants system is also employed as the basis of the resource-guided and agent-based proof planning approach [3], currently under development. Extending the Ω-Ants system this approach also focuses on the cooperation aspect between integrated external reasoners and addresses the question how an agent-based proof planner can be sensibly guided by a resource mechanism.

Acknowledgement

The authors thank John Byrnes for his support in realizing the NIC calculus in Ω-Ants. We furthermore thank S. Autexier, M. Kerber, M. Jamnik, and M. Hübner for fruitful discussions.

References

[1] P. B. Andrews, M. Bishop, S. Issar, D. Nesmith, F. Pfenning, and H. Xi. TPS: A Theorem Proving System for Classical Type Theory. *Journal of Automated Reasoning*, 16(3):321–353, 1996.

[2] C. Benzmüller, J. Byrnes, and V. Sorge. Ω-ANTS for interactive ATP. Unpublished draft:
     www.ags.uni-sb.de/~chris/oants-nic00.ps.gz.

[3] C. Benzmüller, M. Jamnik, M. Kerber, and V. Sorge Towards Concurrent Resource Managed Deduction. Cognitive Science Research Paper CSRP-99-17, University of Birmingham, 1999.

[4] C. Benzmüller and V. Sorge. A Blackboard Architecture for Guiding Interactive Proofs. *Proc. of AIMSA'98*, LNAI 1480, Springer, 1998.

[5] C. Benzmüller and V. Sorge. Critical Agents Supporting Interactive Theorem Proving. *Proc. of EPIA-99*, LNAI 1695, Springer, 1999.

[6] P. Bertoli, J. Calmet, F. Giunchiglia, and K. Homann. Specification and integration of theorem provers and computer algebra systems. *Fundamenta Informaticae*, 39(1–2), 1999.

[7] J. Byrnes. *Proof Search and Normal Forms in Natural Deduction*. PhD thesis, Dep. of Philosophy, CMU, Pittsburgh, PA, USA, 1999.

[8] L. Cheikhrouhou and V. Sorge. *PDS* — A Three-Dimensional Data Structure for Proof Plans. In *Proc. of ACIDCA'2000*, Monastir, Tunisia, 2000.

[9] G. Dowek, Th. Hardin, and C. Kirchner. Theorem proving modulo. Rapport de Recherche 3400, INRIA, France, 1998.

[10] M. Fisher and M. Wooldridge A Logical Approach to the Representation of Societies of Agents. In N. Gilbert and R. Conte, editors, *Artificial Societies*. UCL Press, 1995.

[11] F. Giunchiglia, P. Bertoli, and A. Coglio. The OMRS project: State of the Art. *Electronic Notes in Theoretical Computer Science*, 15, 1998.

[12] F. Giunchiglia, P. Pecchiari, and C. Talcott. Reasoning Theories - Towards an Architecture for Open Mechanized Reasoning Systems. In *Proc. of Frontiers of Combining Systems*, pages 157–174, 1996.

[13] M. J. C. Gordon and T. F. Melham. *Introduction to HOL*. Cambridge University Press, Cambridge, United Kingdom, 1993.

[14] A. Ireland and A. Bundy. Productive Use of Failure in Inductive Proof. *Journal of Automated Reasoning*, 16(1–2):79–111, 1996.

[15] M. Jamnik, M. Kerber, and C. Benzmüller. Towards Learning new Proof Methods in Proof Planning. In this volume.

[16] A. Meier. Übersetzung automatisch erzeugter Beweise auf Faktenebene. Master's thesis, Computer Science Department, Universität des Saarlandes, Germany, 1997.

[17] The ΩMEGA-group. ΩMega: Towards a Mathematical Assistant. *Proc. of CADE–14*, LNAI 1249, Springer, 1997.

[18] S. Owre, S. Rajan, J.M. Rushby, N. Shankar, and M.K. Srivas. PVS: Combining Specification, Proof Checking, and Model Checking. In *Computer-Aided Verification, CAV '96*, LNCS 1102, pages 411–414. Springer, 1996.

[19] V. Sorge. Non-Trivial Computations in Proof Planning. In *Proc. of Frontiers of Combining Systems*, LNCS 1794. Springer, 2000.

The THΞOREM∀ Project: A Progress Report *

Bruno Buchberger   Claudio Dupré   Tudor Jebelean
Franz Kriftner   Koji Nakagawa   Daniela Văsaru
Wolfgang Windsteiger

Abstract. *The Theorema project aims at supporting, within one consistent logic and one coherent software system, the entire mathematical exploration cycle including the phase of proving. In this paper we report on some of the new features of Theorema that have been designed and implemented since the first expository version of Theorema in 1997. In addition, in the conclusion, we formulate design goals for the next version of Theorema.*

1 Introduction

As all the projects in the CALCULEMUS group (see e.g. [1], [15], [6], [7], [10], [2], or [9]), also *Theorema* aims at the integration of computer algebra and theorem proving. The objectives and the design of the *Theorema* system, but also the main differences to other systems that are currently being developed, are described in [5]. In the present paper, we report on the progress made in the *Theorema* project since 1997. Roughly, the new features implemented are:

- the *Theorema* formal text language,
- the *Theorema* computational sessions,
- the Prove-Compute-Solve (PCS) prover of *Theorema*,
- special provers within *Theorema*,
- the cascade-meta-strategy for *Theorema* provers.

*This research is partially supported by the Austrian Science Foundation (FWF-SFB-project F1302) and the Upper Austrian Government (project "PROVE").

98

2   The TH∃OREM∀ Formal Text Language

The core language (expression language) of *Theorema* ([5]) is a version of higher order predicate logic. Expressions like

$$\langle X_i + Y_i \underset{i=1,\ldots,|X|}{\big|} \rangle$$

$$\underset{\underset{\epsilon>0}{\epsilon}}{\forall} \; \underset{\underset{\delta>0}{\delta}}{\exists} \; \underset{\underset{|y-x|<\delta}{y}}{\forall} \; |f[y] - f[x]| < \epsilon$$

are examples of terms and formulae in this language.

However, for composing and manipulating large formal mathematical texts we need to be able to combine the expression language with auxiliary text (labels, key words like "Definition," "Theorem," etc.) and to compose, in a hierarchical way, large mathematical knowledge bases from individual expressions. For this, we designed and implemented the "*Theorema* Formal Text Language."

Here are some typical examples of formal text written in the *Theorema* formal text language:

Definition["continuity", any$[f, x]$,

continuous$[f, x] :\Leftrightarrow \underset{\underset{\epsilon>0}{\epsilon}}{\forall} \; \underset{\underset{\delta>0}{\delta}}{\exists} \; \underset{\underset{|y-x|<\delta}{y}}{\forall} \; |f[y] - f[x]| < \epsilon$ "c2:"]

Definition["fprod", any$[f, g, x]$,

$(f * g)[x] = f[x] * g[x]$ "f*g"]

Proposition["continuity of product", any$[f, x]$,

continuous$[f, x] \wedge$ continuous$[g, x] \Rightarrow$ continuous$[f * g, x]$ "cont*"]

Finally, in order to process knowledge, we provide the "*Theorema* Command Language" in order to *prove* propositions, *compute* values, or *solve* problems, see also Section 2.3.

In the sequel, we will describe the most important features of the "*Theorema* Formal Text Language," which contains three basic categories of Formal Text Elements:

Environments for organizing knowledge in definitions, theorems, etc.,

Built-ins for assigning a "built-in interpretation" to certain symbols, and

Properties for asserting properties of certain operators.

2.1 TH∃OREM∀ Environments

Theorema environments allow to enter knowledge into the system in a style similar to how definitions, propositions etc. are given in mathematical textbooks. Consider a definition like

Definition 1 (Sum of Tuples) *For any two tuples X and Y with $|X| = |Y|$ we define*

$$X \oplus Y := \langle X_i + Y_i \quad | \quad \rangle \qquad (19)$$
$$\qquad\qquad\qquad i=1,\dots,|X|$$

The ingredients of a structure like this are a keyword ("Definition," "Lemma," "Theory," etc. .), a label for later reference ("Sum of Tuples"), one or more mathematical statement(s) (a formula/term or a definition of a new function/predicate), an enumeration of the (free) variables occurring in the statement (X and Y), and, if necessary, conditions on the variables or relations among them. The *Theorema* Formal Text Language supports input of the above definition in the following format:

Definition["Sum of Tuples", any[X, Y], with[$|X| = |Y|$]]
$$\quad X \oplus Y := \langle X_i + Y_i \quad | \quad \rangle \quad "\langle\rangle + \langle\rangle"]$$
$$\qquad\qquad i=1,\dots,|X|$$

More abstract, an environment has the form

Keyword[*env_label*,{any[*vars*] {,with[*cond*]}, }
$$\quad clause_1 \quad \{label_1\} \quad | \quad keyword_1[env\_label_1]$$
$$\qquad\qquad\qquad \{more\ references\}$$

where all fields enclosed in {} are optional and the | denotes an alternative. The user is free to choose any string for *env_label* and the clause labels *label$_i$*. Omitting the clause labels assigns "1," "2," etc. automatically. Labels do not carry any semantics, they are only used for referring to environments and formulae in proofs and computations. The field "any[*vars*]" declares *vars* as (the free) variables. Each variable v in *vars* can carry a type restriction of the form "type[v]" (see also the example in Section 3). The field "with[*cond*]" tells that the variables must satisfy the condition *cond*.

The effect of entering an environment into the system is that the environment is transformed into an internal representation that can be referred to later by *Keyword*[*env_label*]. Knowledge can be grouped using *nested environments*, whose structure is identical except that instead of clauses (formulae with optional labels) there are references to previously defined environments. Typical keywords used for nested environments are "Theory" and "KnowledgeBase."

2.2 Built-ins and Properties

Theorema gives the user full control over the interpretation of any symbol, hence, the automatic interpretation of symbols by the underlying *Mathematica* system must be avoided. For this, the user has the possibility to give implicit knowledge about the interpretation of symbols using the Formal Text Element "**Built-in**." Entering

Built-in["My ops",

$+ \rightarrow$ Times
$* \rightarrow$ Plus]

defines Built-in["My ops"] to translate "+" into the the *Mathematica* built-in function Times and "*" into Plus. In addition, we provide various translations of symbols to their "usual" meaning. Using the Formal Text Element "**Property**" in a similar fashion, it is possible to assert properties of operators (e.g. commutativity of "+"). Each *Theorema* command then provides the possibility to obey implicit knowledge of that kind.

2.3 The TH∃OREM∀ Command Language

In a *Theorema* standard session, the user has maximum control over processing knowledge, i.e. the user can for instance specify an explicit knowledge base, implicit knowledge about operators, or the appropriate prove (simplify, solve) method. Typical *Theorema* commands are:

Prove[Proposition["continuity of product"], by$\rightarrow$PCS,
  using$\rightarrow$ {Definition["continuity"], Definition["fprod"]}],

Compute[$\langle 1, 2, 3 \rangle \oplus \langle 7, 1, -3 \rangle$, using$\rightarrow$ {Definition["Sum of Tuples"],
  built-in$\rightarrow$ {Built-in["Operators"], Built-in["Tuples"]}], or

Compute[$\langle 1, 2, 3 \rangle \oplus \langle 7, 1, -3 \rangle$, using$\rightarrow$ {Definition["Sum of Tuples"],
  built-in$\rightarrow$ {Built-in["My ops"], Built-in["Tuples"]}],

where the options have the following meaning:

using defines the explicit knowledge base to be used.

built-in defines implicit knowledge about symbols used.

by specifies the method to be applied.

  Note the difference in the last two computations: The first one uses "normal" interpretation of symbols provided in Built-in["Operators"] and, thus, results in $\langle 8, 3, 0 \rangle$, whereas the latter employs the user-defined interpretation Built-in["My ops"] from the previous section, thus resulting in $\langle 7, 2, -9 \rangle$.

Moreover, the command language contains administrative commands in order to adjust global settings guarding the system's behavior and to maintain global values. After having set global values appropriately, all options in the calls to "Prove" and "Compute" can be omitted.

3   The TH∃OREM∀ Computational Session

As we saw in the example above, in a *Theorema* standard session, the user interacts with *Theorema* by, first, specifying various definitions, axioms, propositions, and knowledge bases built up from such entities and, then, calls a *Theorema* prover, simplifier, or solver using the *Theorema* Command Language.

The explicit indication of the knowledge base used and the prove (simplify, solve) method applied gives maximum control over the formal development of a mathematical text. This is an important feature of *Theorema*. Typically, current mathematical software systems lack this feature, which is the reason why logically important side-conditions (like conditions on parameters in integrals etc.) cannot be modeled correctly in these systems, see however recent extensions to mathematical software systems like the `Assumptions` option to some commands in *Mathematica* 4 or the `assume` facility in MAPLE.

However, often, the knowledge base used and the method applied is fixed for a long part of a formal text (for example, for an entire section of a book). For such situations, *Theorema* now provides two facilities: One can either define the knowledge base and/or the prove method applied as a global parameter or one can switch to a "computational session." In such a session, it is tacitly assumed that every new definition, axiom, proposition etc. is added to the global knowledge base and that a standard simplifier is applied to the expression entered into an input cell. In other words, in a computational session, *Theorema* behaves very much the same as *Mathematica* or any of the other mathematical software systems. Moreover, the computational session does not need environments as described in Section 2, since there is neither need to refer to individual formulae nor to group them into nested structures.

In order to switch to a computational session, use the command

ComputationalSession[using→Definition["Sum of Tuples"]]

The option `using` gives the opportunity to import knowledge available in the standard session into the global knowledge base for the computational session. Alternatively, one can import this definition using the command

Use[Definition["Sum of Tuples"]]

from inside a computational session. Symbols defined in the standard session are invisible in the computational session unless they are imported. Instead of entering knowledge through environments, definitions can be given directly to the system:

$$\text{any}[\text{is-set}[A, B]]: \quad A \ominus B := \{x \underset{x \in A}{\quad|\quad} x \notin B\}$$

In general, a definition has the form

$\{\text{any}[vars]\{,\text{with}[cond]\}\}$:
$\qquad lhs := rhs$
$\quad \{ more\ definitions \}$

where "any[...]" and "with[...]" have the same semantics like inside an environment in the standard session (note the type specification in the example!). Unlike in a standard session, one can now simply type

$$\{2, 3, 1, 3\} \ominus \{3, 5\}$$

into an input cell of *Mathematica* and evaluate to $\{1, 2\}$ without having to (without being able to!) specify any knowledge base or evaluation method. The knowledge base is the accumulated knowledge built up during the current session, the evaluation method is *Mathematica*'s default expression evaluator. There is no possibility to give implicit knowledge about operator symbols.

4   The Prove-Compute-Solve Prover for Predicate Logic

Many interesting mathematical notions are defined by formulae whose syntactical structure is characterized by a sequence of "alternating quantifiers," i.e. the definitions have the structure

$$p[x, y] \Leftrightarrow \underset{a}{\forall} \underset{b}{\exists} \underset{c}{\forall} \dots q[x, y, a, b, c, \dots]$$

The exploration of theories about notions introduced by such definitions aims at proving, first of all, an arsenal of "rewrite rules" for these notions which later will be helpful in the subsequent proofs of more complicated theorems or for the purpose of "computing examples" involving these notions.

For example, most of the notions introduced in elementary analysis
text books (like the notion of limit, the notion of continuity, the notion of
the growth order of a function etc.) fall into this class. The automated
proof of propositions about such notions is, therefore, a practically important challenge for future mathematical systems, as was pointed out in [4].
Meanwhile, this class of propositions is used by various people as a test set
for their systems, see for example [11].

In *Theorema*, we now implemented a new prover for predicate logic,
which we call the "PCS" ("Prove, Compute, Solve") prover that is particularly suited for proving theorems about notions of the above kind, which, in
this paper, we call "alternating quantifiers theorems." The basic strategy
of this prover simulates what we believe is a frequent and natural strategy used by human provers for routine proofs about alternating quantifiers
theorems.

The "Prove Phase": We first apply all the usual inference rules (in our
particular version of natural deduction, see [5]) of propositional and
predicate logic to the goal and the non-rewrite formulae in the knowledge base until no more such rule can be applied.

The "Compute Phase": Now we use all the rewrite-formulae for simplifying the goal and the formulae of the knowledge base ("computing").
Note that rewrite rules that stem from an implication "expand the
knowledge" but "reduce the goal." Note also that, in this phase,
we allow "semantical pattern matching" (see next paragraph), which
is much stronger than ordinary syntactical pattern matching. The
compute phase may introduce new formulae in the knowledge base
or new goals that can be manipulated as described in the preparation phase and, also, allow again the application of rules of the prove
phase. Hence, we may need to circle through the preparation, prove
and compute phases a couple of times (which, in the exceptional case,
may already yield a proof) before we enter the next phase.

The "Solve Phase": Now we are left with a proof situation in which the
goal has the form

$$\underset{x,y}{\exists}\ G[x, y, ...].$$

This situation essentially specifies a "find problem": We have to find
x^*, y^*, such that $G[x^*, y^*, ...]$ is true under the assumptions collected
in the knowledge base. In this phase, a couple of general rules for

transforming the solve problem are applied and then, depending on the type of the variables $x, y, \ldots$, special solvers are called. For example, if $x, y, \ldots$ are variables ranging over real numbers we call (a full or specialized) version of Collins' cad-method (see [8]) (which is implemented in *Mathematica* version 4.0, see [14]).

"Semantical pattern matching": We explain this idea in an example. Assume that, in the knowledge base, we have the formula

$$\underset{x,y,z,t,\delta,\epsilon}{\forall} (|y * t - x * z| < \delta * (\epsilon + |z| + 1) + |x| * \epsilon \Longleftarrow$$
$$(|y - x| < \delta \wedge |t - z| < \epsilon))$$

and the proof goal contains

$$|f_0[y] * g_0[y] - f_0[x_0] * g_0[x_0]| < \epsilon_0.$$

Then, by syntactical rewriting, the goal cannot be reduced because, ϵ_0 cannot be obtained from $\delta * (\epsilon + |z| + 1) + |x| * \epsilon$ by a substitution. However, in this situation, the goal can be reduced to

$$\underset{\delta,\epsilon}{\exists} (\delta * (\epsilon + |g_0[x_0]| + 1) + |f_0[x_0]| * \epsilon = \epsilon_0 \wedge$$
$$|f_0[y] - f_0[x_0]| < \delta \wedge |g_0[y] - g_0[x_0]| < \epsilon).$$

It turns out that the PCS method is quite powerful for generating, with almost no superfluous search, natural (easy to understand) proofs for many elementary "alternating quantifiers theorems." (Note, however, that these "elementary" theorems give lots of headache to beginning students of mathematics. Also, they are outside the scope of both purely algebraic provers like Collins' cad-method and the usual general predicate logic provers. Thus, being able to generate natural proofs for these theorems, in our view, is a definite step forward.)

We demonstrate the method by showing the proof generated by the PCS prover of *Theorema* for the proposition introduced in Section 2. In fact, the actual notebook generated contains "nested proof cells" that can be used to browse the proof in a "structured way," as was explained in [5]. Note that the entire proof including all intermediate natural language text (i.e. everything between the two horizontal lines below) is generated completely automatically by the PCS prover:

Prove:

(cont*) $\displaystyle\mathop{\forall}_{f,g,x}$ (continuous $[f,x] \wedge$ continuous $[g,x] \Rightarrow$ continuous $[f * g, x])$

under the assumptions:

(c2:) $\displaystyle\mathop{\forall}_{f,x}$ continuous $[f,x] \Leftrightarrow \mathop{\forall}_{\substack{\epsilon\\ \epsilon>0}} \mathop{\exists}_{\substack{\delta\\ \delta>0}} \mathop{\forall}_{\substack{y\\ |y-x|<\delta}} (|f[y] - f[x]| < \epsilon),$

(f*g) $\displaystyle\mathop{\forall}_{f,g,x} ((f * g)[x] = f[x] * g[x]),$

(dist*) $\displaystyle\mathop{\forall}_{x,y,z,t,\delta,\epsilon} (|y * t - x * z| < \delta * (\epsilon + |z| + 1) + |x| * \epsilon \Leftarrow$
$(|y - x| < \delta \wedge |t - z| < \epsilon)),$

(min>) $\displaystyle\mathop{\forall}_{m,M1,M2} (\min[M1, M2] > m \Leftrightarrow M1 > m \wedge M2 > m),$

(<min) $\displaystyle\mathop{\forall}_{m,M1,M2} (m < M1 \wedge m < M2 \Leftrightarrow m < \min[M1, M2]).$

We assume

(1)   continuous $[f_0, x_0] \wedge$ continuous $[g_0, x_0],$

and show

(2) continuous $[f_0 * g_0, x_0].$

Formula (1.1), by (c2:), implies

(3) $\displaystyle\mathop{\forall}_{\substack{\epsilon\\ \epsilon>0}} \mathop{\exists}_{\substack{\delta\\ \delta>0}} \mathop{\forall}_{\substack{y\\ |y-x_0|<\delta}} (|f_0[y] - f_0[x_0]| < \epsilon).$

By (3) we can introduce a Skolem function such that

(4) $\displaystyle\mathop{\forall}_{\substack{\epsilon\\ \epsilon>0}} \delta_0[\epsilon] > 0 \wedge \mathop{\forall}_{\substack{\epsilon\\ \epsilon>0}} \mathop{\forall}_{\substack{y\\ |y-x_0|<\delta_0[\epsilon]}} (|f_0[y] - f_0[x_0]| < \epsilon).$

Formula (1.2), by (c2:), implies

(5) $\displaystyle\mathop{\forall}_{\substack{\epsilon\\ \epsilon>0}} \mathop{\exists}_{\substack{\delta\\ \delta>0}} \mathop{\forall}_{\substack{y\\ |y-x_0|<\delta}} (|g_0[y] - g_0[x_0]| < \epsilon).$

By (5) we can introduce a Skolem function such that

(6) $\displaystyle\mathop{\forall}_{\substack{\epsilon\\ \epsilon>0}} \delta_1[\epsilon] > 0 \wedge \mathop{\forall}_{\substack{\epsilon\\ \epsilon>0}} \mathop{\forall}_{\substack{y\\ |y-x_0|<\delta_1[\epsilon]}} (|g_0[y] - g_0[x_0]| < \epsilon).$

Formula (2), using (c2:), is implied by

(9) $\displaystyle\mathop{\forall}_{\substack{\epsilon\\ \epsilon>0}} \mathop{\exists}_{\substack{\delta\\ \delta>0}} \mathop{\forall}_{\substack{y\\ |y-x_0|<\delta}} (|(f_0 * g_0)[y] - (f_0 * g_0)[x_0]| < \epsilon).$

We assume

(10) $\epsilon_0 > 0,$

and show

(11) $\exists\limits_{\substack{\delta \\ \delta>0}} \forall\limits_{\substack{y \\ |y-x_0|<\delta}} (|(f_0 * g_0)[y] - (f_0 * g_0)[x_0]| < \epsilon_0).$

We have to find $\delta_2{}^*$ such that

(12) $\delta_2{}^* > 0 \wedge \forall\limits_{\substack{y \\ |y-x_0|<\delta_2{}^*}} (|(f_0 * g_0)[y] - (f_0 * g_0)[x_0]| < \epsilon_0).$

(IN THE FULL PROOF TEXT, IN A COUPLE OF EASY REDUCTION STEPS, THIS PROOF PROBLEM IS FURTHER REDUCED TO THE PROBLEM)

We have to find $\delta_2{}^*$, $\delta_3{}^*$, and $\epsilon_1{}^*$ such that)

(19) $(\delta_3{}^* * (\epsilon_1{}^* + |g_0[x_0]| + 1) + |f_0[x_0]| * \epsilon_1{}^* = \epsilon_0) \wedge$
$\delta_3{}^* > 0 \wedge \epsilon_1{}^* > 0 \wedge (\delta_2{}^* = \min[\delta_0[\delta_3{}^*], \delta_1[\epsilon_1{}^*]]).$

Summarizing, we reduced the proof to a solving problem. We have to find $\delta_2{}^*$, $\delta_3{}^*$, $\epsilon_1{}^*$ such that (19) holds under the current knowledge. The problem can be solved by calling the *Mathematica* implementation of the Cylindrical Algebraic Decomposition Algorithm. Hence, we are done. The solution is of the form

$0 < \delta_3{}^* < \frac{\epsilon_0}{1+|g_0[x_0]|}$, $\epsilon_1{}^* = \frac{\epsilon_0-(1+|g_0[x_0]|)*\delta_3{}^*}{\delta_3{}^*+|f_0[x_0]|}$, $\delta_2{}^* = \min[\delta_0[\delta_3{}^*], \delta_1[\epsilon_1{}^*]].$

In fact, the above proof generated by the *Theorema* PCS prover produces much more detailed information on the "solving terms" than is normally done in proofs by human mathematicians. This information is quite interesting and would make it possible to formulate a much stronger version of the proposition, namely a version in which the dependence of the objects to be found (like $\epsilon_1^*, \delta_1^*, \delta_3^*$) on the objects given (like ϵ_0) is explicit. Also, the above proof follows the style of proving, in which, during the generation of the proof, full motivation is given why certain constructions are done in the way displayed. This style of presenting proofs is often considered to be "more pedagogical" than the style where the constructions are presented "by the teacher" without any motivation and the student afterwards is left (with the easy) task of just verifying that the constructions are appropriate.

However, as soon as this detailed and explicit version of the proof is generated it would be possible (by a "proof simplification" algorithm) to translate the proof into a version which suppresses the details of the construction and/or re-arranges the proof into a style where the constructions are just given and then verified.

In a similar fashion, a set theory prover will be available soon. This prover will use additional inference rules from set theory in the prove and compute phase in order to translate proof situations involving notions from set theory into proof situations in predicate logic.

5 Special Provers Within *Theorema*: Gröbner Bases Prover and Gosper-Zeilberger Prover

Special provers (solvers, and simplifiers) can be be integrated into *Theorema*. A special prover P for a theory (i.e. a collection of formulae) T is a prover satisfying the "Correctness Meta-Theorem for P w.r.t. T":

> For all knowledge bases K and goals G, if P produces a proof of G under the assumption K, then G is a logical consequence of $K \bigcup T$.

In fact, some of the algebraic provers we are currently interested in, like the Gröbner bases prover and Collins' prover, are also complete. Similarly, the correctness of special solvers and simplifiers for a given theory T is defined.

Typically, the reduction of a proof problem to a strong special prover (or solver) is a process that deserves explanation in a proof text whereas the actual call of the special prover (e.g. the computation of a Gröbner basis) is not something the user is interested to see in detail. Rather, one normally prefers to use these provers in a "black box" style. We give two examples: The automated proof of universally quantified boolean combinations of equalities over the complex numbers using the Gröbner bases method and the automated proof of combinatorial identities using the Gosper-Zeilberger prover with the extension of this method by Paule and Schorn, see [13]. Both methods are now accessible by calls from within *Theorema*, i.e. the proof problems can be formulated in the *Theorema* formal language, the reduction of the proof problems to the black box methods is explained by natural language text in the proofs generated and the actual black box computation are then just presented by their result.

Since the reduction of a *proof problem* given as a boolean combination of polynomial equalities and inequalities to a non-linear multivariate polynomial *solve problem* is already well-documented in the literature, see for example [3], we only show an example of the Gosper-Zeilberger method. Note that, actually, this method is more than a prover, it is a theorem generator (or, in other words, a simplifier): For a given summation term, it produces the "closed form", i.e. a simplified term or recurrence. The example we give, shows the power of the method: It produces a closed form for a sum whose evaluation was proposed as a SIAM REVIEW problem, see [12].

Formula["SIAM series",

$$\sum_{k=1}^{n} \frac{(-1)^{k+1}(4k+1)(2k)!}{2^k(2k-1)(k+1)!2^k k!} \ ]$$

`Prove[` Formula[`"SIAM series"`], by→GosperZeilbergerProver].

(Note that, again, the entire proof including all intermediate natural language text is generated completely automatically by the prover.)

Theorem: If $-1 + n$ is a natural number, then:

$$\sum_{k=1}^{n} \left(\frac{-(-1)^k 2^{-2k}(2k)!(1+4k)}{k!(1+k)!(-1+2k)} \right) = 1 + \frac{-(-1)^n 2^{-2n}(2n)!}{n!(1+n)!}.$$

Proof:
Let Δ_k denote the forward difference operator in k. Then the Theorem follows from summing the equation

$$\frac{-(-1)^k 2^{-2k}(2k)!(1+4k)}{k!(1+k)!(-1+2k)} = \Delta_k \left[\frac{(-1)^k 2^{1+-2k}(2k)!(1+k)}{k!(1+k)!(-1+2k)} \right]$$

over the range $k = 1, \dots, n$.

The equation is routinely verifiable by dividing the right-hand side by the left-hand side and simplifying the resulting rational function:

$$\frac{\frac{(-1)^{1+k} 2^{1+-2(1+k)}(2(1+k))!(2+k)}{(1+k)!(2+k)!(-1+2(1+k))} - \frac{(-1)^k 2^{1+-2k}(2k)!(1+k)}{k!(1+k)!(-1+2k)}}{\frac{-(-1)^k 2^{-2k}(2k)!(1+4k)}{k!(1+k)!(-1+2k)}}$$

to 1.

6   Extending Existing Provers by Meta-Strategies

Given a prover P, one can apply various strategies for enhancing the proving power of P. One of these strategies is what we call the "cascade": Intuitively, the idea is that, given a goal G and a knowledge base K, we let P try to find a proof. If P succeeds, we stop and present the proof. If not, we let a "failure analyzer" analyze the proof attempt and conjecture a lemma L, which could be strong enough to allow P to prove G from $K \bigcup L$. Now we let P try to prove L from K.

If it succeeds we let P try, again, to prove G but this time under the assumption $K \bigcup L$, otherwise we let the failure analyzer work on the failing proof. More formally, given a prover P and a "conjecture from failure generator" C, the following recursive "cascade" may result in a much stronger prover that, in fact, does not only prove more theorems than P but, on the way of proving a goal from a knowledge base, gradually extends the knowledge base by "useful" lemmas:

Cascade[G,K,C,P]:=
   proof-attempt=Prove [G,K,P];
   if proof-attempt is successful,
        then Return[{ "proved",K }];
        else L=C[proof-attempt];
   {proof-value,new-K }=Cascade[L,K,C,P];
   if proof-value="proved",
        then Cascade[G,K $\bigcup$ L,C,P];
        else
            if new-K=K,
                then Return[{ "failed",K }];
                else Cascade[G,new-K,C,P].

We show the effect of the cascade in the case of a simple induction prover for natural numbers and a simple conjecture generator that conjectures a generalized equality over the natural numbers from a special instance of the equality, which occurs in a failing proof.

Starting from the definition

Definition["Addition",

$$\underset{m}{\forall}\ m + 0 = 0 \qquad\qquad \underset{m,n}{\forall}\ m + n^+ = (m + n)^+\ ]$$

we might want to prove

Proposition["Commutativity +",

$$\underset{m,n}{\forall}\ m + n = n + m\ ].$$

This can be tried by calling

Prove[Proposition["Commutativity +"], using→Definition["Addition"], by→NNEqIndProver]].

With this simple prover, the proof will fail: It generates a proof attempt that is stuck at the situation where it should prove (for a constant m_1)

$$(\text{F}) \qquad\qquad (0 + m_1)^+ = m_1^+$$

A human reader would immediately conjecture that, maybe, 0 is also a left unit, i.e.

$$(\text{L}) \qquad\qquad \underset{m}{\forall}\ 0 + m = m$$

and that, maybe, if this conjecture is true, the proof of commutativity could go on and could be completed. Producing the conjecture (L) from the failure line is a relatively easy process: "Strip off all identical outer symbols from the left-hand and right-side terms of (F) and turn the constants into variables." With this simple procedure, which we programmed and called ConjectureGenerator, the call of the above cascade

Prove[Proposition["Commutativity +"], using→Definition["Addition"],
    by→Cascade[NNEqIndProver,ConjectureGenerator]]

produces, successively and without any further user-interaction, five note-books that contain the following proofs and proof attempts:

1) A failing proof attempt for proving the goal $\underset{m,n}{\forall}\ m + n = n + m$

   from the knowledge base

   $\underset{m}{\forall}\ m + 0 = m,\ \underset{m,n}{\forall}\ m + n^+ = (m + n)^+.$

2) A successful proof for the goal $\underset{m}{\forall}\ 0 + m = m$

   (which is automatically generated from analyzing the failing proof 1)
   from the knowledge base

   $\underset{m}{\forall}\ m + 0 = m,\ \underset{m,n}{\forall}\ m + n^+ = (m + n)^+.$

3) A failing proof attempt for proving the goal $\underset{m,n}{\forall}\ m + n = n + m$

   from the knowledge base

   $\underset{m}{\forall}\ 0 + m = m,\ \underset{m}{\forall}\ m + 0 = m,\ \underset{m,n}{\forall}\ m + n^+ = (m + n)^+.$

4) A successful proof for the goal $\underset{n,m}{\forall}\ n^+ + m = (n + m)^+$

   (which is automatically generated from analyzing the failing proof 3)
   from the knowledge base

   $\underset{m}{\forall}\ 0 + m = m,\ \underset{m}{\forall}\ m + 0 = m,\ \underset{m,n}{\forall}\ m + n^+ = (m + n)^+.$

5) A successful proof for the goal $\underset{m,n}{\forall}\ m + n = n + m$

   from the knowledge base

   $\underset{n,m}{\forall}\ n^+ + m = (n + m)^+,\ \underset{m}{\forall}\ 0 + m = m,$
   $\underset{m}{\forall}\ m + 0 = m,\ \underset{m,n}{\forall}\ m + n^+ = (m + n)^+.$

Note that, in addition to being able to prove the original goal, the combination of NNEqIndProver with ConjectureGenerator in the cascade also automatically produces two additional, and in fact quite natural and interesting, lemmata.

7 Conclusion

We reported on some of the new features of *Theorema*. The next steps in the *Theorema* project will mainly concentrate on improving the PCS prover(s) and implementing and integrating various special provers described in the literature that have proven to be particularly successful. *Theorema*, in its present version (Version 1), offers a fixed arsenal of provers (solvers, and simplifiers). We are also working on a major re-design of *Theorema* (Version 2) that will make it possible for the user to formulate his own provers (solvers, and simplifiers) in a language that will be particularly suited for this task, i.e. will make it particularly easy to program provers (solvers, and simplifiers) that, in addition to the abstract proof objects, also produce natural language intermediate explanatory proof texts.

References

[1] C. Benzmüller, L. Cheikhrouhou, D. Fehrer, A. Fiedler, X. Huang, M. Kerber, M. Kohlhase, K. Konrad, E. Melis, A. Meier, W. Schaarschmidt, J. Siekmann, and V. Sorge. ΩMEGA: Towards a Mathematical Assistant. In *Proceedings of the 14th Conference on Automated Deduction* (Townsville, Australia, 1997), W. McCune, Ed., no. 1249 in LNAI, Springer-Verlag, pp. 252–255. See also http://www.ags.uni-sb.de/projects/deduktion/.

[2] P. G. Bertoli, J. Calmet, F. Giunchiglia, and K. Homann. Specification and Integration of Theorem Provers and Computer Algebra Systems. *Fundamenta Informaticae* (1999). Accepted for publication.

[3] B. Buchberger. Applications of Gröbner bases in Non-Linear Computational Geometry. In *Proc. Workshop on Scientific Software* (IMA, Minneapolis, USA, 1987), Springer.

[4] B. Buchberger. The Objectives of the *Theorema* Project. Talk at the 1996 CALCULEMUS Meeting, University of Rome, Italy, 1996.

[5] B. Buchberger, T. Jebelean, F. Kriftner, M. Marin, E. Tomuţa, and D. Văsaru. A Survey of the *Theorema* Project. In *Proceedings of ISSAC'97 (International Symposium on Symbolic and Algebraic Computation)* (Maui, 1997), W. Kuechlin, Ed., ACM Press, pp. 384–391.

[6] A. Bundy, F. van Harmelen, C. Horn, and A. Smaill. The Oyster-Clam system. In *Proceedings of the 10th International Conference on Automated Deduction* (1990), M. E. Stickel, Ed., Springer-Verlag, pp. 647–648.

[7] O. Caprotti, and A. M. Cohen. Integrating Computational and Deduction Systems Using *OpenMath*. In *Proceedings of CAL-CULEMUS: Systems for Integrated Computation and Deduction* (Trento, Italy, 1999), A. Armando and T. Jebelean, Eds., Electronics Notes in Theoretical Computer Science, Elsevier. See also http://www.nag.co.uk/projects/openmath/omsoc/.

[8] G. E. Collins. Quantifier Elimination for Real Closed Fields by Cylindrical Algebraic Decomposition. In *Second GI Conference on Authomata Theory and Formal Languages* (1975), vol. 33 of *LNCS*, Springer-Verlag, Berlin, pp. 134–183.

[9] R. L. Constable et al. *Implementing Mathematics with the Nuprl Proof Development System*. Prentice-Hall, 1986. See also http://www.cs.cornell.edu/Info/Projects/NuPrl.

[10] K. Homann and J. Calmet. Structures for Symbolic Mathematical Reasoning and Computation. In *Proceedings of the International Symposium on Design and Implementation of Symbolic Computation Systems (DISCO 96)* (1996), J. Calmet and C. Limongelli, Eds., no. 1128 in LNCS, Springer, pp. 217–228.

[11] E. Melis. The "Limit" Domain. In *Proceedings of AIPS-98* (1998), R. Simmons, M. Veloso, and S. Smith, Eds.

[12] P. Paule. Problem 94-2. *SIAM REVIEW* (1995), 105–106.

[13] P. Paule and M. Schorn. A Mathematica Version of Zeilberger's Algorithm for Proving Binomial Coefficient Identities. *J. Symbolic Computation 20* (1995), 673–698.

[14] A. Strzebonski. Solving Equations and Inequalities with *Mathematica*. In *Proceedings of the* Mathematica *Developer Conference* (1999).

[15] A. Trybulec and H. Blair. Computer Assisted Reasoning with MIZAR. In *Proceedings of the 9th International Joint Conference on Artificial Intelligence* (Los Angeles, CA, Aug. 1985), A. Joshi, Ed., Morgan Kaufmann, pp. 26–28. See also http://mizar.uw.bialystok.pl/.

How to Formally and Efficiently Prove Prime(2999)

Olga Caprotti*   Martijn Oostdijk

Abstract. *This paper focuses on how to use Pocklington's criterion to produce efficient formal proof-objects for showing primality of large positive numbers. First, we describe a formal development of Pocklington's criterion, done using the proof assistant* COQ. *Then we present an algorithm in which computer algebra software is employed as oracle to the proof assistant to generate the necessary witnesses for applying the criterion. Finally, we discuss the implementation of this approach and tackle the proof of primality for some of the largest natural numbers expressible in* COQ.

1   Introduction

The problem of showing whether a positive number is prime or composite is historically recognized to be an important and useful problem in arithmetic. Since Euclid's times, the interest in prime numbers has only been growing. For today's applications, primality testing is central to public key cryptography and for this reason is still heavily investigated in number theory [10].

Although the problem is clearly decidable, the trivial algorithm that checks for every number q such that $q \leq \sqrt{n}$ whether $q|n$, is far too inefficient for practical purposes. There exist several alternative methods to check primality and in this paper we deal with a classical criterion due to Pocklington in 1914 [9]. Our interest is motivated by the fact that in order to produce a proof of primality the criterion needs to find numbers that verify certain algebraic equalities. These numbers are easily generated using a computer algebra package, for instance GAP [8].

---

*This author was supported by the OpenMath Esprit project 24969.

This paper presents how Pocklington's criterion can be employed to produce efficient formal proof-objects that show primality of large positive numbers in a proof assistant such as COQ [5]. This entails a formal development in COQ of Pocklington's criterion, the study of how computer algebra software can assist as oracle to COQ, and finally details about the implementation of these ideas.

Notice that the cooperation of theorem provers with computer algebra is essential for being able to solve this task. Theorem provers are very limited on the amount and type of computations they can perform [1], however, they are very well suited to organizing the logical steps of a proof. On the other hand, although computer algebra systems have algorithms for deciding whether a number is prime or not, it is hard for them to produce a proof of primality. Thus, the winning strategy is to combine both kinds of system.

The structure of this paper is as follows. The formalization of Pocklington's criterion is described in Section 2. Section 3 gives the architectural details for using it to produce formal proofs of primality with the aid of computer algebra oracles. Our implementation of the algorithm is discussed in Section 4 and to conclude the paper, timings for some benchmarks are summarized in Section 5.

2 Formalization

The Pocklington criterion is one of many number theoretical results that are useful for verifying primality of a positive number n. The work we present in this section stems from the study that Elbers did in his PhD thesis [7] and is a continuation of [3]. The complete development consists of COQ vernacular files and is available online [6].

Lemma 1 (Pocklington's Criterion) *Let* $n \in \mathrm{N}$, $n > 1$ *with* $n - 1 = q \cdot m$ *such that* $q = q_1 \cdots q_t$ *for certain primes* $q_1, \ldots, q_t$. *Suppose that* $a \in \mathrm{Z}$ *satisfies* $a^{n-1} = 1 \,(\mathrm{mod}\ n)$ *and* $\gcd(a^{\frac{n-1}{q_i}} - 1, n) = 1$ *for all* $i = 1, \ldots, t$. *If* $q \geq \sqrt{n}$, *then* n *is a prime.*

Proof:

Let $p | n$ and Prime(p), put $b = a^m$.
Then $b^q = a^{mq} = a^{n-1} = 1 \,(\mathrm{mod}\ n)$.
So $b^q = 1 \,(\mathrm{mod}\ p)$.
Now q is the order of b in Z_p^*, because:

Suppose $b^{\frac{q}{q_i}} = 1 \,(\mathrm{mod}\ p)$, then $a^{\frac{mq}{q_i}} = a^{\frac{n-1}{q_i}} = 1 \,(\mathrm{mod}\ p)$.

There exist $\alpha, \beta \in \mathbb{Z}$ such that $\alpha(a^{\frac{n-1}{q_i}} - 1) + \beta n = 1 \pmod{p}$.
So, $\alpha(1-1) + \beta 0 = 1 \pmod{p}$. Contradiction.
By Fermat's little theorem: $b^{p-1} = 1 \pmod{p}$,
therefore $q \leq p - 1$, so $\sqrt{n} \leq q < p$.
Hence for every prime divisor p of n: $p > \sqrt{n}$.
Therefore $\mathsf{Prime}(n)$.

$\square$

Although the proof is easy from a mathematical viewpoint, formalization in CoQ is not straightforward at all.

Definitions. The first step in formalizing this proof in CoQ, is to find all the mathematical notions that are used in the proof. Although the CoQ system is really just a logical system, many notions such as the natural numbers, the integers, relations, lists, etc. are defined in the standard library. Still, many notions that are part of the repertoire of any mathematician, like division and primality, are not (yet) in the standard library. One has to define these concepts first. Here are some examples of predicates that are needed to formulate and prove the criterion.

$$
\begin{aligned}
\mathsf{Divides}(n, m) &= \exists q : \mathrm{N}.(m = n * q) \\
\mathsf{Prime}(n) &= (n > 1) \wedge \forall q : \mathrm{N}.(\mathsf{Divides}(q, n) \to (q = 1 \vee q = n)) \\
\mathsf{Mod}(a, b, n) &= \exists q : \mathrm{Z}.(a = b + n \cdot q) \\
\mathsf{Order}(b, q, p) &= (0 < q) \wedge \mathsf{Mod}(b^q, 1, p) \wedge \\
&\quad \forall d : \mathrm{N}.(((0 < d) \wedge \mathsf{Mod}(b^d, 1, p)) \to (q < d))
\end{aligned}
$$

For brevity $\mathsf{Divides}(n, m)$ will be denoted by $n|m$, and $\mathsf{Mod}(a, b, n)$ by $(a = b \pmod{n})$. Examples of other concepts that need to be defined are modulo arithmetic, the order of an element in a finite group, exponentiation, and greatest common divisor.

Lemmas. The second step in formalizing an informal proof in CoQ, is to divide it into smaller lemmas. In the case of Pocklington's proof this is necessary for three reasons. First of all the proof takes rather large steps. Look for example at line 4 of the informal proof. To prove that q is the order of b in Z_p^*, denoted as $\mathsf{Order}(b, q, p)$, a contradiction is derived from the assumption that $b^{\frac{q}{q_i}} = 1 \pmod{p}$ for some prime factor q_i of q. This is based on a number of non-trivial technical lemmas:

order_ex: $\forall p: \mathrm{N}.\,\mathsf{Prime}(p) \to \forall b: \mathrm{Z}.\exists d: \mathrm{N}.(\mathsf{Order}(b,d,p))$
order_div: $\forall b: \mathrm{Z}.\forall x: \mathrm{N}.\forall p: \mathrm{N}.\mathsf{Order}(b,x,p) \to$
$$\forall y: \mathrm{N}.(y > 0) \wedge (b^y = 1 \pmod p)) \to x|y$$
tlemma3: $\forall a, b: \mathrm{N}.(0 < a < b) \wedge (a|b) \to$
$$\exists q_i : \mathrm{N}.((q_i|b) \wedge \mathsf{Prime}(q_i) \wedge (a|\tfrac{b}{q_i}))$$

The second reason why we need to divide the proof into smaller lemmas is that the standard library of CoQ does not include many of the concepts and propositions needed to prove Pocklington. After these concepts are defined, many trivial lemmas about them are needed before they can be used in the informal proof. For example in line 3 of the proof it is stated that the equality $b^q = 1 \pmod n$ implies the equality $b^q = 1 \pmod p$. From a mathematical viewpoint this is a trivial step, but CoQ requires us to prove a lemma about modulo arithmetic first.

modpq_modp:
$$\forall a, b: \mathrm{Z}.\forall p, q: \mathrm{N}.(a = b \pmod{pq}) \to (a = b \pmod p)$$

The third reason for dividing the proof into smaller lemmas is that the informal proof uses *forward reasoning* whereas the CoQ system, being a goal directed theorem prover, uses *backward reasoning*. The user is presented with a goal to prove and works his way back to the assumptions on which this goal depends by means of tactics. The informal proof, however, starts by introducing an arbitrary prime divisor p of n. The underlying (technical) lemma is

primepropdiv:
$$\forall n: \mathrm{N}.(n > 1) \wedge (\forall p: \mathrm{N}.\,\mathsf{Prime}(p) \wedge p|n \to (p > \sqrt{n})) \to \mathsf{Prime}(n)$$

Once this lemma is proved, one can apply it and the current goal $\mathsf{Prime}(n)$ will be replaced by a new obligation to prove that $n > 1$ and for all prime divisors p of n that $p > \sqrt{n}$. The application of the lemma corresponds to lines 1,10, and 11 of the proof of Lemma 1.

Theories.   Now that we know what lemmas to prove, we can distinguish between those that express properties of the mathematical concepts that were introduced in a previous step, and technical ones which correspond to high level steps in the informal proof. Lemmas of the former category are organized together with the relevant definitions in *theories*. These can be reused in the formalization of other theorems. In formalizing Pockling-ton's theorem, for example, a theory about prime numbers was developed,

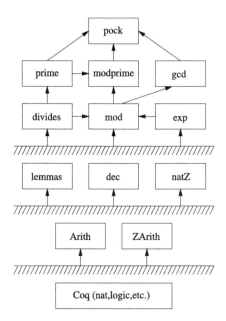

Figure 1. Theories and their dependencies.

containing the definition of the **Prime** predicate and some lemmas, among which **primepropdiv** mentioned earlier. Figure 1 depicts the different theories used to formalize the proof of Pocklington's theorem.

Some of these theories deal with mathematics on a meta level. The *dec* theory contains some lemmas that are useful for proving decidability of predicates in general. Decidability in the context of constructive theorem provers like Coq means that the principle of the excluded middle holds. The formalization of Pocklington's criterion described here is done constructively using the lemmas in *dec*. Alternatively the principle of the excluded middle could have been assumed.

Other examples of meta results are lemmas to switch from the Z type to the N type, and lemmas to switch from quantifiers ∀ and ∃ to finite conjunctions and disjunctions.

Technical lemmas. The final step of the formalization is proving the high-level technical lemmas and the theorem itself. If the base of lemmas in the theories is large enough, this should be straightforward.

3   Architecture

In this section we describe how Pocklington's Criterion can be used to produce a formal and efficient primality proof for a relatively big[1] prime number. In a skeptical approach one invokes an outside oracle to supply the theorem prover with the necessary witnesses for applying Pocklington's criterion. For instance, computer algebra systems can act as oracles when algebraic equalities have to be verified. For the skeptical approach to work, a computer algebra system must be able to supply both a fast decision for the primality of a positive number n and in the affirmative case additional extra information for building a proof-object.

First of all, as described in Section 2, we have formally proved Pocklington's Criterion using CoQ. This implies that the formal proof of primality of a positive number n can be reduced to the proof of the necessary conditions to Pocklington's criterion. The algorithm for building a CoQ proof-object for Prime(n) can be summarized by the steps outlined in Figure 2.

Interaction with computer algebra oracles takes place mostly in Step (1) of the algorithm. Figure 3 depicts the flow of information input and output between the algorithm and the computer algebra package in a specific instance. When the computer algebra software is given a positive number n, it first tests whether the number is indeed prime. If not, it returns back the number. If the number n is prime, then the system can easily compute the numbers a, q and m as follows. For a take the primitive root (mod n), namely an element a such that $a^{n-1} = 1, \pmod{n}$ and $a^i \neq 1, \pmod{n}$ for $i = 1, \ldots, n-2$. For q, consider the prime factorization $n - 1 = q_1 \ldots q_t \ldots q_k$, where $q_1 \geq \ldots \geq q_t \geq \ldots \geq q_k$, and take $q = q_1 \ldots q_t$ for the smallest t such that $q \geq \sqrt{n}$. Finally, for m take $m = (n-1)/q$. All the operations to compute the appropriate a, $q = q_1 \ldots q_t$, and m are carried out by the computer algebra package upon receiving the prime number n. Notice that these witnesses, computed as described, satisfy the hypotheses of Pocklington's criterion. Subgoals (4)(a) and (4)(c) are clearly true. Condition (4)(b) is true because n is prime and $\gcd(a^{\frac{n-1}{q_i}} - 1, n)$ cannot be n. If it was n, then $a^{\frac{n-1}{q_i}} = 1 \pmod{n}$ for an exponent $\frac{n-1}{q_i} < n - 1$. However, this is not possible because a is the primitive root (mod n).

The computer algebra oracle is also called in Step (4)(b). It computes the coefficients for the linear combinations generated by the gcd proof obligations using a straightforward extension of the Euclidean gcd algorithm. Most computer algebra systems provide this algorithm as primitive. The

---

[1] See the next section for a discussion on the size of the prime.

$$PocklingtonC\,(n;T)$$

Input: n a prime number.

Output: T a tactic script for proving primality of n by Pocklington's criterion.

(1) [Find witnesses.]
    Let a be the primitive root mod n. Choose q and m such that $n = qm+1$, $q \geq 0$, $m \geq 0$. Compute the prime factorization of q in $q_1 \cdot \ldots \cdot q_t$

(2) [Recursion Step]
    Apply recursively $PocklingtonC\,(q_i;S_i)$ for $i = 1,\ldots,t$ to every prime factor q_i in the factorization of q, thus obtaining tactic scripts $S_1,\ldots,S_t$.

(3) [Apply Pocklington's]
    Apply Pocklington's criterion using the parameters a, q, q_1, ..., q_t, and m in order to prove $\mathsf{Prime}(n)$.

(4) [Prove the subgoals]
    Provide the tactic scripts S_a, S_b, and S_c for proving the subgoals corresponding to the hypotheses of Pocklington's criterion.

    (a) $a^{n-1} = 1(\mathrm{mod}\ n)$ is shown by by a divide and conquer strategy in which the exponent gets smaller until the computation is trivial.

    (b) $\gcd(a^{\frac{n-1}{q_i}} - 1, n) = 1$, $i = 1,\ldots,t$ is shown by proving that 1 is a linear combination of $a^{\frac{n-1}{q_i}} - 1$ and n (mod n).

    (c) $n \leq q^2$ is shown trivially.

(3) [Output] Assemble the tactic scripts S_1, ..., S_t, S_a, S_b, and S_c in the tactic script T for proving $\mathsf{Prime}(n)$.

Figure 2. Pocklington Criterion Algorithm.

oracle also computes the result of the exponentiations for a^{n-1} (mod n) and $a^{\frac{n-1}{q_i}}$ and for all the intermediate steps in the divide and conquer procedure.

To summarize the overall picture, the algorithm $PocklingtonC$ can be used to produce a CoQ tactic script that generates a proof-object for the primality of a positive number. The only requirements on the computer algebra systems used as oracles is the ability to perform integer computa-

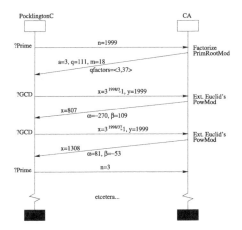

Figure 3. Communication of witnesses for $n = 1999$.

tions like prime testing, factorization, gcd computation and some modular arithmetic. If the communication uses the standard *OpenMath* [4], then the architecture allows for multiple computer algebra oracles, see Figure 4. All responses of COQ can be predicted, so the communication to COQ can be in one direction only.

4 Implementation

Our implementation of the architecture described above consists of a JAVA application in which the user enters a positive integer n and selects a computer algebra package running on a remote server. If the number is prime, the computer algebra package is repeatedly invoked for a concrete value of n and for the subsequent recursive calls of the factors. It computes the witnesses according to the message sequence chart in Figure 3. The application then generates a COQ tactic script that can be automatically communicated to COQ when proving the goal Prime(n). COQ returns a verified proof-object which is displayed by the applet. In a different version of the implementation, the user can prompt for the automatic generation of an interactive document describing the formal proof [2].

In practicality, the algorithm outlined in Section 3 has to take into account limitations on the size of the prime number n. Computer algebra software, like GAP, is able to test primality for integers that are up to 13 digits long. For bigger integers, the primality test are probabilistic and re-

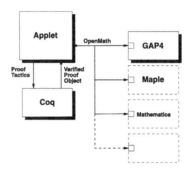

Figure 4. Overall Architecture.

turn a *probable prime*. For instance, testing numbers with several hundreds digits is quite feasible in GAP4 using `IsPrimeInt` or `IsProbablyPrimeInt`[2]. Concerning factorization, `FactorsInt` is guaranteed to find all factors less than 10^6 and will find most factors less than 10^{10}.

A more serious limitation on the size of n is imposed by theorem proving software like COQ, which is not geared to handling big integers or big natural numbers. To partially overcome this limitation, we have formalized and proved a variant of Pocklington's theorem based on the integer library `ZArith`, which uses a binary representation for numbers, but this is not yet enough if we want to prove primality for really big numbers.

5   Benchmarks

The maximum prime number that can be checked by COQ is rather limited by the fact that our primality predicate is defined on the natural numbers. COQ's natural numbers `nat` are inductively defined using a zero and a successor constructor. The largest natural number we were able to use in COQ 6.3.1 is 9343.

The `ZArith` library provides a more efficient datatype Z for storing numbers. This datatype is used to do the necessary computations to verify the witnesses. The maximum number representable in this datatype is much higher, but an attempt to prove Prime(9343) still resulted in a COQ stack overflow.

We tested the generated tactic scripts for all primes between 2 and 2999 and measured the run time needed by COQ to produce and check the

---

[2]However, for such titanic primes we are unable to check the generated proofs.

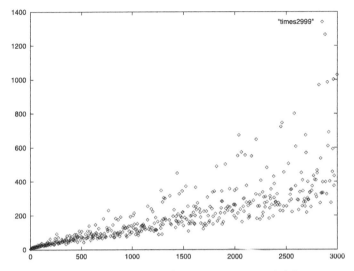

Figure 5. Time, in seconds, needed by CoQ to verify Prime(n) for $n = 3 \ldots 2999$.

proof-object on a Sparc Ultra Enterprise with 2GB of memory. The results are plotted in Figure 5. Obviously the general trend is that larger primes need more time. However, some numbers are much harder due to an unfortunate prime factorization of q. For example when proving Prime(2039), the algorithm is forced to choose $q = 1019$, since the prime factorization of 2038 is $2 \cdot 1019$. Now, 1028 in its turn has as prime factorization $2 \cdot 509$. In the end recursive calls for 3, 7, 127, 509, and 1019 are needed to prove Prime(2039), and it takes about 674 seconds to verify the proof In contrast to verify the proof generated for $n = 2939$ only needs recursive calls for 3, 7, and 113 and only takes about 275 seconds.

6   Conclusions

We have shown how by combining computer algebra oracles and theorem provers it is possible to automatically produce proofs of primality that are efficiently and formally verifiable. The primality proofs are obtained according to Pocklington's criterion.

To formally prove Pocklington's criterion in CoQ, we needed to first prove about 150 smaller lemmas taking approximately 4000 lines of code. No matter how enjoyable, this is still a lot of hard work. A less skeptical approach that assumes many of the needed lemmas as axioms requires less effort and might still produce proofs that are acceptable in other communities.

The architecture for using Pocklington's criterion relies on computer algebra oracles. We interpret these oracles as mathematical servers providing computational capabilities on the network and the JAVA application that produces the tactic script as a client to these servers. In this general view, our experiments are an example of how to use computer algebra in theorem proving and an investigation on the tools that are required to effectively carry out the integration. We profited greatly from our work in using the standard communication language *OpenMath* to interface to a variety of symbolic computation systems.

The final implementation we produced, although able to prove primality of 2999, is far from being usable for proving primality of interestingly big numbers. We are able to generate the tactic script for such big numbers, however the theorem prover is not able to deal with the large datastructures involved.

References

[1] H. Barendregt and E. Barendsen. Efficient Computations in Formal Proofs. To Appear.

[2] O. Caprotti and M. Oostdijk. Proofs in Interactive Mathematical Documents. In *Proceedings of AISC '2000, Artificial Intelligence and Symbolic Computation, Theory, Implementations and Applications*, July 2000. To appear.

[3] Olga Caprotti and Arjeh Cohen. Integrating Computational and Deduction Systems Using OpenMath. In *Proceedings of Calculemus 99*, Trento, July 1999.

[4] The OpenMath Consortium. The OpenMath Standard. OpenMath Deliverable 1.3.3a, OpenMath Esprit Consortium, http://www.nag.co.uk/projects/OpenMath.html, August 1999. O. Caprotti, D. P. Carlisle and A. M. Cohen Eds.

[5] Projet Coq. *The Coq Proof Assistant: The standard library*, version 6.3-1 edition. Available at http://www.ens-lyon.fr/LIP/groupes/coq.

[6] Pocklington development files. http://crystal.win.tue.nl/~olga/openmath/pocklington/.

[7]   Hugo Elbers. *Connecting Informal and Formal Mathematics*. PhD thesis, Eindhoven University of Technology, Eindhoven, The Netherlands, May 1998.

[8]   Martin Schönert et al. *GAP 4 manual*. Available at `http://www-history.mcs.st-and.ac.uk/gap/`.

[9]   H. C. Pocklington. The determination of the prime or composite nature of large numbers by fermat's theorem. *Proc. Cambridge Philosophical Society*, 18(6):29–30, 1914.

[10] Paulo Ribenboim. *The New Book of Prime Number Records*. Springer Verlag, 1996.

On the EA-Style Integrated Processing of Self-Contained Mathematical Texts

Anatoli I. Degtyarev   Alexander V. Lyaletski

Marina K. Morokhovets

Abstract. *In this paper[1], we continue to develop our approach to theorem proof search in the EA-style, that is theorem proving in the framework of integrated processing mathematical texts written in a 1st-order formal language close to the natural language used in mathematical papers. This framework enables constructing a sound and complete goal-oriented sequent-type calculus with "large-block" inference rules. In particular, it contains the formal analogs of such natural proof search techniques as definition handling and auxiliary proposition application. The calculus allows to incorporate symbolic computations in an inference search.*

1   Introduction

In early 1960s, V.Glushkov initiated the programme of work on the automation of theorem proof search in mathematics. That programme received later the name "Evidence Algorithm" (EA, or AO due to Russian "Algoritm Ochevidnosti"). One of the basic notions in the EA programme was the notion of evidence of a proof step which changes as EA progresses.

"...The structure of Evidence Algorithm needs to leave room for unbounded replenishment of EA with new and new blocks for the purpose of creating more levels of hierarchy. For practical application of EA, it is important to achieve such a level in its progress when an average length of a proof (including refutation examples construction) comes practically to an average length of proofs required in textbooks and monographs, and then in

---

[1]This research was partially supported by the project INTAS 96-0760 and by the EPSRC grant GR/M46631 given to the first author.

special papers. Therewith, of course, besides Evidence Algorithm proper, information base of the system which contains descriptions (in practical mathematical logic language) of different kinds of concepts used in a specific mathematical theory under consideration, as, also, properties of these concepts, procedures for the generation and investigation of examples, etc is to be developed. All this information abundance ought to be used by EA just as a human does it." [1]

In accordance with the principles of this programme (see, for example, [2]), the following scheme of automated theorem proving in the EA-style have been developed. An assertion to be proven is immersed in a mathematical text written in so-called Theory Language (TL) [3]. The language TL was designed to meet the following requirements: to be a communication link between a user and a computer within an automated theorem proving system; to be a formal language for representing and storing mathematical data; to be a high-level language close to a language of usual mathematical publications. The language TL contains a rich collection of linguistic structures that, on the one hand, have precise definitions (in terms of BNF), and, on the other hand, are similar to constructions of natural languages. That is why it has been selected as the first approximation to a family of formalized mathematical languages for EA. Then, such a TL-text is transformed into a so-called $TL1$-text, if it is possible. A $TL1$-text consists of $TL1$-sentences, which are, on the one hand, analogs of 1st-order logic formulas, and, on the other hand, preserve the signature of an original TL-text, its syntax and structure (i.e. partitioning into definition sections, auxiliary proposition sections and a theorem to be proved). In what follows, the dual nature of a $TL1$-sentence will be often exploited. Now a $TL1$-text is a source for proof search in the EA-style.

In the framework of such an approach to automated theorem proving in the EA-style, a special sequent-type formalism has been developed [4]. As a result, two so-called a-sequent calculi have been suggested [5]. They are sound and complete and satisfy the following requirements: syntactical form of an initial problem should be preserved; deduction should be done in the signature of initial texts, in particular, preliminary skolemization should be non-obligatory; proof searching should be goal-oriented; equality handling should be separated from deductive processes.

In this paper, we continue to develop our approach to theorem proof search in the EA-style, that is theorem proving in the framework of integrated processing mathematical texts written in a formal 1st-order language close to the natural language used in mathematical papers. This framework enables constructing a sound and complete goal-oriented sequent-type calculus with "large-block" inference rules. In particular, it contains the formal analogs of such natural proof search techniques as definition handling and auxiliary proposition application.

1.1   Basic Computer-oriented Approaches to Logical Inference Search in 1st-order Classical Calculi

The beginning of EA programme realization can be referred to the publication of the paper [6] on the heuristic procedure for theorem proof search in Group Theory in which there was made allowance (on a formal level) for some proof search methods used in mathematical papers. Then that formal technique was extended to specific fragments of Set Theory. Its final completion appeared as the specific calculus AGS (Auxiliary Goal Search) [7] which was meant for ascertaining the validity of 1st-order classical logic formulas. In parallel, research on a formal language [8] oriented to the representation of mathematical texts intended to be processed by a computer was performed. That language can be viewed as an analog of a classical 1st-order language enriched with means which enable to use constructions and notation more convenient and usual for mathematicians.

Historically, it is worth noting that the end of the 1950s – beginning of the 1960s can be characterized as the period of coming into existence computers with such a performance rate, information capacity, and flexibility that programming complex intellectual processes became feasible. As a response to the emergence of computing machinery of that sort a series of papers appeared in which the issues of the implementation of Gentzen-type calculi [9] and inference search methods relying on the results of Skolem [10] and Herbrand [11] were discussed. For more detail, the reader is referred to, for example, [12, 13, 14, 15], etc. It might be well to point out that in those first papers an answer to the principal question about a possibility to use computers for mathematical reasoning was provided. But the lack of machine-oriented techniques for the optimization of enumeration during reasoning impeded getting a proof even for rather simple true assertions in 1st-order classical logic. Investigations with the aim of improving the efficiency of the proof search methods proposed resulted in coming into existence Kanger's calculus [16] (of Gentzen type) and Robinson's resolution method [17] (of Skolem-Herbrand type). For those days, the latter yielded the most efficient machine-oriented inference search technique for the 1st-order classical predicate calculus by using Robinson's unification algorithm. (S.Yu.Maslov's inverse method [18] is worth of special noting here. It can be formulated in resolution terms for 1st-order classical logic, although the initially proposed scheme of the inverse method was subsequently extended to non-classical logics.) Thus studies on automated theorem proving were later on concentrated mainly upon improving the potentialities of the Skolem-Herbrand approach. So, unification algorithms which took account of specific features of a particular 1st-order theory with equality were proposed (A-unification, AC-unification, etc.). The ways of building-in special equality handling rules (for instance, the paramodulation rule) into resolution-based methods have been investigated. Subsequent attempts for the advancement of the Skolem-Herbrand approach resulted in arising tableaux methods, the connection graph method, goal-oriented search methods, etc.

The lower efficiency of Kanger's approach (as compared with the Skolem-Herbrand approach) can be explained by the fact that Kanger-type calculi (and the calculus AGS is among them) do not need the obligatory carrying out of preliminary skolemization that can result in arising the superfluous enumeration caused by the possibility of different orders of logical (mainly, quantifier) inference rules application and the necessity of center formulas duplication when applying some of those rules.

At the same time, machine-oriented Gentzen-Kanger-type methods reflect proof techniques which are more "natural" for a human. They also enable to construct rather flexible tools to support a dialogue between a user and a computer during interactive inference search, and to facilitate understanding a proof process by a user. So, in the framework of EA programme realization activity, achieving the improved search efficiency of AGS by a possibility both of "transition" to the sound (and complete, if possible) deductive processing of mathematical texts written in a formalized mathematical language close to usual mathematical publication languages and the application of mathematical facts gained came forward as a central problem of proof search automation.

As a result of research performed, the sound and complete 1st-order calculus of a-sequents with an original notion of an admissible substitution has been suggested [4]. This notion has been introduced with the aim of the optimization of additional efforts connected with the possibilities of different orders of quantifier elimination without preliminary skolemization. (It has been shown later that the notion of an admissible substitution can be easily "built-in" into standard Gentzen calculi [19].)

Along with Gentzen-Kanger-type calculi, Skolem-Herbrand-type calculi have been investigated [20]. The retrospective point of view on the linguistic tools and deductive systems of EA can be found in [21].

1.2   The Current State of the EA-style Inference Search

Nowadays we carry out research work within the EA programme at a new level of understanding the problem of automated theorem proving and taking into account existing trends in the development of program systems which support "doing mathematics."

In [5], the a-sequent calculi (gS and mS) were proposed, and the formal description of gS was given, but the calculus mS was only illustrated by an example, whereas the inference rules of mS were not presented. To make up this deficiency, we give in this paper the formal description of mS. This paper reflects continuing efforts to attack the problem of automated theorem-proving in the EA-style by the application of definitions and auxiliary propositions. Following [5], we denote the corresponding calculus as mS. To make the paper "self-contained" enough we give here the necessary definitions.

The calculus mS permits to present an initial problem as a text in a certain 1st-order formal language containing definitions and auxiliary propositions, and to use analogs of such natural theorem proving techniques as the application of definitions and auxiliary propositions. The peculiarity of our approach is that needed definitions and auxiliary propositions are extracted from a self-contained mathematical text written in the formal language TL [3] approximated to languages of usual mathematical papers. A self-contained mathematical text is a text that, in addition to a proposition to be proved, also includes assumptions, propositions, and definitions that can be used when the proof of a given assertion is searching for.

Processing a self-contained mathematical text for the purpose of proving a given theorem is divided into three parts:

(1) writing down an original mathematical text as a TL-text;

(2) translating the TL-text into a $TL1$-text;

(3) searching for a proof in mS within the $TL1$-text environment.

2   The Calculus mS and Theorem Proving

In this section, we present the calculus mS as a deductive basis for solving the problem of the validity of a given assertion in the context of a natural mathematical text. After the text is written in TL-language and converted into a $TL1$-text, a theorem proof search is carried out using the inference rules of mS.

The calculus mS is an extension of gS [5] with additional inference rules for the application of definitions and auxiliary propositions.

2.1   Preliminaries

The basic object of mS is an a-sequent. An a-sequent may be considered as a special generalization of the standard notion of a sequent. We consider a-sequents having one object ("goal") in its succedent only.

We treat here 1st-order classical logic in the form of the sequent calculus G given in [22].

We treat the notion of a substitution as in [17]. Any substitution component is considered to be of the form t/x, where x is a variable and t is a term of a substitution.

Let L be a literal, then $\sim L$ denotes its complement. We use the expression $L(t_1, \ldots , t_n)$ to denote that $t_1,...,t_n$ is a list of all the terms (possibly, with repetitions) occupying the argument places in the literal L in the order of their occurrences in L. If x, y are variables and F is a formula then $F|_y^x$ denotes the result of replacing x with y.

We also assume that besides usual variables there are two countable sets of special variables, namely unknown variables and fixed variables (dummies and parameters in the terminology of [16]).

An ordered triple $< w, F, E >$ is called *an ensemble* iff w is a sequence (a word) of unknown and fixed variables, F is a 1st-order formula, and E is a set of pairs of terms t_1, t_2 (equations of the form $t_1 = t_2$).

An a-sequent is an expression of the form $[\mathcal{P}], [\mathcal{D}], [B], < w_1, P_1, E_1 >,...,$ $< w_n, P_n, E_n > \Rightarrow < w, F, E >$, where $< w_1, P_1, E_>,...,< w_n, P_n, E_n >$, $< w, F, E >$ are ensembles, $[B]$ is a list of literals, possibly empty, $[\mathcal{P}]$ and $[\mathcal{D}]$ are lists of $TL1$-sentences corresponding to auxiliary propositions and definitions, respectively.

Ensembles in the antecedent of an a-sequent are called premises, and an ensemble in the succedent of an a-sequent is called a goal of this a-sequent. The collection of the premises is thought as a set. So, the order of the premises is immaterial.

Let W be a set of sequences of unknown and fixed variables, and s be a substitution. Put $A(W, s) = \{< z, t, w >: z$ is a variable of s, t is a term of s, $w \in W$, and z lies in w to the left of some fixed variable from $t\}$. Then s is said to be *admissible* for W iff (1) the variables of s are unknown variables only, and (2) there are not (different) elements $< z_1, t_1, w_1 >,...,< z_n, t_n, w_n >$ in $A(W, s)$ such that $t_2/z_1 \in s,...,t_n/z_{n-1} \in s, t_1/z_n \in s$ $(n > 0)$.

Decomposition of some $TL1$-sentence F by its principal logical connective and possible interaction with P_i results in generating new a-sequents. The sets $E_1,...,E_n$, E define the terms to be substituted for the unknown variables in order to transform every equation $t_1 = t_2$ from $E_1,...,E_n, E$ to identity $t = t$ after applying to $E_1,...,E_n, E$ a substitution chosen in a certain way. The sets $w_1,...,w_n, w$ serve to check whether the substitutions generated during proof searching are admissible. Note, that in any a-sequent some (or all) sequences from $w_1,...,w_n, w$ and some (or all) sets from $E_1,...,E_n, E$ may be empty.

An initial a-sequent is constructed as follows. Suppose, a self-contained $TL1$-text Txt consists of the collection $[\mathcal{D}]$ of definitions and the collection $[\mathcal{P}]$ of auxiliary propositions, and a theorem T is given, which can be interpreted in terms of the calculus G as a sequent of the form $P_1, \ldots, P_n \Rightarrow F$. Then an a-sequent $[\mathcal{P}], [\mathcal{D}], [], <, P_1, >, \ldots, <, P_n, > \Rightarrow <, F, >$ will be considered as an initial a-sequent (w.r.t. T and Txt).

During proof searching in mS an inference tree is constructed. At the beginning of a search process it consists of an initial a-sequent. The subsequent nodes of the inference tree are generated in accordance with the rules described below. Inference trees grow "from top to bottom".

2.2   The Calculus mS

In the formulation of rules below, M denotes a set of premises, $[\mathcal{D}]$ ($[\mathcal{D}_1]$, $[\mathcal{D}_2]$) is a list of definitions, possibly empty, $[\mathcal{P}]$ ($[\mathcal{P}_1]$, $[\mathcal{P}_2]$) is a list of auxiliary propositions, possibly empty. Sometimes, $[\mathcal{P}]$, $[\mathcal{D}]$ will be omitted when they are not used.

Let us introduce inductively a notion of *a positive (negative) occurrence of a literal L in a formula F (denoted by $F\lfloor L^+ \rfloor$ and $F\lfloor L^- \rfloor$, respectively) modulo equations* in a rigorous way:
(I) suppose that a literal F ($\sim F$) can be obtained from $L(t_1, \ldots, t_n)$ by means of replacing $t_1,...,t_n$ with some terms $t'_1,...,t'_n$. Then L is said to have a positive (negative) occurrence in the literal F modulo the equations $t_1 = t'_1,...,t_n = t'_n$;
(II.1) if $F\lfloor L^+ \rfloor$ ($F\lfloor L^- \rfloor$) modulo the equations $t_1 = t'_1, \ldots, t_n = t'_n$ and F_1 is a formula then L has a positive (negative) occurrence (modulo the equations $t_1 = t'_1, \ldots, t_n = t'_n$) in the following formulas: $F \wedge F_1$, $F_1 \wedge F$, $F \vee F_1$, $F_1 \vee F$, $F_1 \supset F$, $\forall x F$, $\exists x F$;
(II.2) if $F\lfloor L^+ \rfloor$ ($F\lfloor L^- \rfloor$) modulo the equations $t_1 = t'_1, \ldots, t_n = t'_n$ and F_1 is a formula then L has a negative (positive) occurrence (modulo the equations $t_1 = t'_1, \ldots, t_n = t'_n$) in the following formulas: $F \supset F_1$, $\neg F$;
(III) there are no other cases of positive (negative) occurrences of L in F.

Goal Splitting Rules (GS)

The rules GS are used for the elimination of the principal logical connective from the $TL1$-sentence in the goal of an a-sequent processed. The application of any rule results in generation of a new a-sequent with only one goal (and, possibly, with new premises). The elimination of the $TL1$-equivalents of proposition connectives is done according to 1st-order classical logic (it can be easily expressed in the terms of derivative rules of standard Gentzen-type calculi [22]), and $w_1,...,w_n, w, E_1,...,E_n$ therewith are not changed. Essential deviation from traditional Gentzen inference search techniques is observed in the processing of quantifiers. This deviation reflects specific quantifier handling techniques investigated in [4], where variables of eliminated quantifiers are replaced by undefined or fixed variables depending on an eliminated quantifier. Therewith w, but not $w_1,...,w_n, E_1,...,E_n, E$, is changed, and new premises can be generated.

Propositional Rules

$(\Rightarrow \supset_1)$-rule:

$$\frac{[B], M \Rightarrow < w, F \supset F_1, E >}{[B], M, < w, F, E > \Rightarrow < w, F_1, E >}$$

$(\Rightarrow \supset_2)$-rule:

$$\frac{[B], M \Rightarrow < w, F \supset F_1, E >}{[B], M, < w, \neg F_1, E > \Rightarrow < w, \neg F, E >}$$

$(\Rightarrow \wedge)$-rule:

$$\frac{[B], M \Rightarrow < w, F \wedge F_1, E >}{[B], M \Rightarrow < w, F, E > \quad [B], M \Rightarrow < w, F_1, E >}$$

$(\Rightarrow \vee_1)$-rule:

$$\frac{[B], M \Rightarrow < w, F \vee F_1, E >}{[B], M, < w, \neg F, E > \Rightarrow < w, F_1, E >}$$

$(\Rightarrow \vee_2)$-rule:

$$\frac{[B], M \Rightarrow < w, F \vee F_1, E >}{[B], M, < w, \neg F_1, E > \Rightarrow < w, F, E >}$$

$(\Rightarrow \neg)$-rule:

$$\frac{[B], M \Rightarrow < w, \neg F, E >}{[B], M \Rightarrow < w, F', E >}$$

where F' is the result of one-step transferring "$\neg$" into F.

Quantifier Rules

$(\Rightarrow \forall)$-rule:

$$\frac{[B], M \Rightarrow < w, \forall x F, E >}{[B], M \Rightarrow < w\bar{x}, F|_{\bar{x}}^{x}, E >}$$

where $\bar{x}$ is a new fixed variable.

$(\Rightarrow \exists)$-rule:

$$\frac{[B], M \Rightarrow < w, \exists x F, E >}{[B], M, < w, \forall x \neg F, E > \Rightarrow < wx', F|_{x'}^{x}, E >}$$

where x' is a new unknown variable.

Auxiliary Goal Rules (AG)

The rules AG reflect the fact that mS is oriented to proof searching by certain transformation of the sentences of self-contained mathematical $TL1$-texts. In terms of Gentzen-type calculi, the rules AG can be interpreted as the elimination of principal logical connectives from $TL1$-sentences, which occur in premises. (Those sentences are generated deterministically beginning with such a premise $< w_i, P_i, E_i >$ that P_i contains a positive occurrence (modulo equations) of the $TL1$-sentence F from the goal of an input a-sequent for AG. As to the elimination of principal logical connectives in premises, the remarks referring to GS are true, excluding, naturally, remarks on $w_1,...,w_n, w, E_1,...,E_n, E$.) The application of AG results in the generation of m ($m > 0$) a-sequents with new goals $< w'_1, F_1, E'_1 >, \ldots, < w'_m, F_m, E'_m >$ and, possibly, some new (w.r.t. the input a-sequent for AG) premises.

Propositional Rules

$(\supset_1 \Rightarrow)$-rule:

$$\frac{[B], < w, F\lfloor L^- \rfloor \supset F_1, E' >, M \Rightarrow < w', L, E >}{[B], < w, (\neg F)\lfloor L^+ \rfloor, E' >, M \Rightarrow < w', L, E > \quad [B, \sim L], M \Rightarrow < w, \neg F_1, E' >}$$

$(\supset_2 \Rightarrow)$-rule:

$$\frac{[B], < w, F \supset F_1\lfloor L^+ \rfloor, E' >, M \Rightarrow < w', L, E >}{[B], < w, F_1\lfloor L^+ \rfloor, E' >, M \Rightarrow < w', L, E > \quad [B, \sim L], M \Rightarrow < w, F, E' >}$$

$(\vee_1 \Rightarrow)$-rule:

$$\frac{[B], < w, F \vee F_1\lfloor L^+ \rfloor, E' >, M \Rightarrow < w', L, E >}{[B], < w, F_1\lfloor L^+ \rfloor, E' >, M \Rightarrow < w', L, E > \quad [B, \sim L], M \Rightarrow < w, \neg F, E' >}$$

$(\vee_2 \Rightarrow)$-rule:

$$\frac{[B], < w, F\lfloor L^+ \rfloor \vee F_1, E' >, M \Rightarrow < w', L, E >}{[B], < w, F\lfloor L^+ \rfloor, E' >, M \Rightarrow < w', L, E > \quad [B, \sim L], M \Rightarrow < w, \neg F_1, E' >}$$

$(\wedge_1 \Rightarrow)$-rule:

$$\frac{[B], < w, F\lfloor L^+ \rfloor \wedge F_1, E' >, M \Rightarrow < w', L, E >}{[B], < w, F\lfloor L^+ \rfloor, E' >, < w, F_1, E' >, M \Rightarrow < w', L, E >}$$

$(\wedge_2 \Rightarrow)$-rule:

$$\frac{[B], < w, F \wedge F_1\lfloor L^+ \rfloor, E' >, M \Rightarrow < w', L, E >}{[B], < w, F, E' >, < w, F_1\lfloor L^+ \rfloor, E' >, M \Rightarrow < w', L, E >}$$

$(\neg \Rightarrow)$-rule:

$$\frac{[B], < w, \neg(F\lfloor L^- \rfloor), E' >, M \Rightarrow < w', L, E >}{[B], < w, F'\lfloor L^+ \rfloor, E' >, M \Rightarrow < w', L, E >}$$

where F' is the result of one-step transferring "$\neg$" into F.

Termination Rules

$(\Rightarrow \sharp_1)$-rule:

$$\frac{[B], < w, L(t_1, \ldots, t_n), E' >, M \Rightarrow < w', L(t'_1, \ldots, t'_n), E >}{M \Rightarrow < w, \sharp, E'' >}$$

(Here $E'' = E' \cup E \cup \{t_1 = t'_1, \ldots, t_n = t'_n\}$; $L(t_1, \ldots, t_n)$, $L(t'_1, \ldots, t'_n)$ are literals.)

$(\Rightarrow \sharp_2)$-rule:

$$\frac{[B_1, L(t_1, \ldots, t_n), B_2], M \Rightarrow < w', L(t'_1, \ldots, t'_n), E >}{[B_1, L(t_1, \ldots, t_n), B_2], M \Rightarrow < w', \sharp, E' >}$$

(Here $E' = E \cup \{t_1 = t'_1, \ldots, t_n = t'_n\}$; $L(t_1, \ldots, t_n)$, $L(t'_1, \ldots, t'_n)$ are literals.)

Quantifier Rules

$(\forall \Rightarrow)$-rule:

$$\frac{[B], < w, \forall x(F\lfloor L^+ \rfloor), E' >, M \Rightarrow < w', L, E >}{[B], < wx', F|^x_{x'}\lfloor L^+ \rfloor, E' >, < w, \forall x F, E' >, M \Rightarrow < w', L, E >}$$

where x' is a new unknown variable.

$(\exists \Rightarrow)$-rule:

$$\frac{[B], < w, \exists x(F\lfloor L^+ \rfloor), E' >, M \Rightarrow < w', L, E >}{[B], < w\overline{x}, F|^x_{\overline{x}}\lfloor L^+ \rfloor, E' >, M \Rightarrow < w', L, E >}$$

where $\overline{x}$ is a new fixed variable.

Definition Application Rule and Auxiliary Proposition Rule

Structuring $TL1$-texts according to substantive sections (i.e. definitions, propositions, etc.) enables introducing in mS the *definition application rule* (DA) and the *auxiliary proposition rule* (AP) in a natural way. These rules can be viewed as specific variants of AG. They represent analogs of natural theorem-proving techniques for the application of definitions and auxiliary propositions.

The rule DA is formulated in view of the structure of the definition section of a TL-text. According to the syntax of the language TL, a definition section consists of two parts, namely, a description part (or description) and a definition part. A description part is a collection of assumptions which satisfies the following closure condition: for any variable x that has an occurrence in some assumption from the description there exists an assumption (in the same description) of the form "Let x be K" or "$x \in M$," where K is a concept, M is a term. A definition part is a TL-sentence of the form $A(x_1, \dots, x_n)$ IFF $\mathcal{F}(x_1, \dots, x_n)$, where $x_1, \dots, x_n$ are variables and every x_i ($i = 1, \dots, n$) has an occurrence in the description part. A formula $A(x_1, \dots, x_n)$, as a rule, is an atom, and $\mathcal{F}(x_1, \dots, x_n)$ is a formula that does not include the atom A (if we restrict ourselves to non-recursive definitions).

A definition section with a description part consisting of the assumptions $F_1, \dots, F_k$ and with a definition part of the form $A(x_1, \dots, x_n)$ IFF $\mathcal{F}(x_1, \dots, x_n)$ can be viewed as the formula $\forall x_1 \dots x_n (F_1 \wedge \dots \wedge F_k \supset (A(x_1, \dots, x_n) \equiv \mathcal{F}(x_1, \dots, x_n)))$. Since the application of a definition in a mathematical practice is generally done as if one of the formulas, $\forall x_1 \dots x_n (F_1 \wedge \dots \wedge F_k \supset (A(x_1, \dots, x_n) \supset \mathcal{F}(x_1, \dots, x_n)))$ or $\forall x_1 \dots x_n (F_1 \wedge \dots \wedge F_k \supset (\mathcal{F}(x_1, \dots, x_n) \supset A(x_1, \dots, x_n)))$, were used, we treat a definition section as a pair of the above formulas.

An auxiliary proposition is a TL-text section consisting of a collection of assumptions and a conclusion. The conclusion can be viewed as a formula $F(x_1, \dots, x_n)$, where $x_1, \dots, x_n$ are variables occurring in the assumptions of the proposition. The collection of assumptions satisfies the above closure condition or it can be transformed in such a way to satisfy the closure condition (we do not consider this transformation here, see, for example [23], [20] on this issue).

In DA-rule below, $F\lfloor L^+ \rfloor$ denotes either $(A(x_1, \dots, x_n) \supset \mathcal{F}(x_1, \dots, x_n))$ or $(\mathcal{F}(x_1, \dots, x_n) \supset A(x_1, \dots, x_n))$.

DA-rule:

$$\frac{[\mathcal{P}], [\mathcal{D}_1, \forall x_1 \dots x_n (F_1 \wedge \dots \wedge F_k \supset F\lfloor L^+ \rfloor), \mathcal{D}_2], [B], M \Rightarrow < w', L, E >}{\mathcal{A}_1 \Rightarrow < w', L, E > \quad \mathcal{A}_1' \Rightarrow < x_1' \dots x_n', F_1 \wedge \dots \wedge F_k, >}$$

(Here $\mathcal{A}_1$ denotes $[\mathcal{P}], [\mathcal{D}_1, \forall x_1 \dots x_n (F_1 \wedge \dots \wedge F_k \supset F\lfloor L^+ \rfloor), \mathcal{D}_2], [B], < x_1' \dots x_n', F\lfloor L^+ \rfloor, >, M$; $\mathcal{A}_1'$ denotes $[\mathcal{P}], [\mathcal{D}_1, \forall x_1 \dots x_n (F_1 \wedge \dots \wedge F_k \supset F\lfloor L^+ \rfloor), \mathcal{D}_2], [B, \sim L], M$.)

AP-rule:

$$\frac{[\mathcal{P}_1, \forall x_1 \dots x_n (F_1 \wedge \dots \wedge F_k \supset F\lfloor L^+ \rfloor), \mathcal{P}_2], [\mathcal{D}], [B], M \Rightarrow < w', L, E >}{\mathcal{A}_2 \Rightarrow < w', L, E > \quad \mathcal{A}_2' \Rightarrow < x_1' \dots x_n', F_1 \wedge \dots \wedge F_k, >}$$

(Here $\mathcal{A}_2$ denotes $[\mathcal{P}_1, \forall x_1 \dots x_n (F_1 \wedge \dots \wedge F_k \supset F\lfloor L^+ \rfloor), \mathcal{P}_2], [\mathcal{D}], [B], < x_1' \dots x_n', F\lfloor L^+ \rfloor), >, M$; $\mathcal{A}_2'$ denotes $[\mathcal{P}_1, \forall x_1 \dots x_n (F_1 \wedge \dots \wedge F_k \supset F\lfloor L^+ \rfloor, \mathcal{P}_2], [\mathcal{D}], [B, \sim L], M$.)

Premise Addition Rule (PA)

The rule PA affects the whole proof search tree. After every application of $(\forall \Rightarrow)$-rule $((\exists \Rightarrow)$-rule$)$, the new premise $< wx', F|_{x'}^{x}\lfloor L^{+}\rfloor, E' > (< w\overline{x},$ $F|_{\overline{x}}^{x}\lfloor L^{+}\rfloor, E' >)$ is added to antecedents of all a-sequents containing a premise with a formula which includes the marked occurrence of $< w,$ $\forall x(F\lfloor L^{+}\rfloor), E' > (< w, \exists x(F\lfloor L^{+}\rfloor), E' >)$ through the current tree.

Axioms

Axioms are a-sequents of the form $[\mathcal{P}], [\mathcal{D}], [B], M \Rightarrow < w, \natural, E >$, where $\natural$ denotes an empty formula.

An Inference Tree

The assertion T to be proved is represented as a substantive $TL1$-section "theorem," in which conditions, or assumptions, and a conclusion are separated, and an initial a-sequent (with respect to T) is constructed with the assumptions and conclusion in its antecedent and succedent, respectively. (The remained part Txt of the $TL1$-text is given as the set of definitions and auxiliary propositions.)

Applying the rules "from top to bottom" to the input a-sequent and afterwards to its "heirs," and so on, we finally obtain an *inference tree* (w.r.t. the theorem T to be proved and "environmental" $TL1$-text Txt). The inference tree Tr is called a *proof tree* for an input a-sequent if and only if (1) every leaf of Tr is an axiom; (2) there exists the unifier s of all equations from Tr; (3) s is admissible (in the sense of this paper) for the set of all sequences of fixed and unknown variables from the leaves of Tr.

At any moment during inference search, it is possible to test whether a current inference tree can be transformed into a proof tree. When the construction of a proof tree is made in an interactive mode, a user may initiate this test. For testing, techniques for finding the most general unifier can be used.

3 Main Results

It was noted above that any $TL1$-sentence can be treated as an analog of some 1st-order classical logic formula. It enables constructing formula patterns of such units of a $TL1$-text as the theorem to be proved, a definition, an auxiliary proposition and to treat a self-contained $TL1$-text as a set of 1st-order formulas. So, it is possible to understand unambiguously such terms as "$TL1$-text consistency," "logical consequence of a theorem from a given $TL1$-text," and "validity" (of the theorem to be proved) without special defining the semantics of the $TL1$-language. With this in mind, we state main results about mS as follows.

Proposition 1 (soundness and completeness of mS). $TL1$-theorem T is a logical consequence of a consistent $TL1$-text Txt (which does not include T) if and only if a proof tree (with an initial sequent w.r.t. T and Txt) can be constructed in mS.

Proposition 2. A $TL1$-theorem T is valid if and only if a proof tree (with an initial sequent w.r.t. T only) can be constructed in mS.

We note, as a side-result, that rather rich collection of rules in mS enables to construct various proof search strategies which model proofs from usual mathematical texts, and, by maintaining the interactive mode of proof search, to allow a user to influence a proof process actively. If such a strategy (with or without participation of a human) ensures an exhaustive search, then propositions 1 and 2 guarantee the soundness and completeness of the strategy.

4   Conclusion

The described approach to logical inference search enables to gain the following benefits.

1. Through using an original notion of an a-sequent instead of a standard notion of a sequent it is possible to develop a quantifier handling technique (without preliminary skolemization) that allows achieving inference search efficiency comparable with the efficiency of methods requiring preliminary skolemization.

2. A-sequents allow to reduce, by rather standard logical transformation, the assertion to be proven to a number of new auxiliary assertions without specifying which terms should be substituted instead of variables.

3. Forming the collections of equations (which can be treated as certain constraints) during proof search allows to postpone finding a solution for equations up to arbitrary moment of time, to separate it from deductive process, and then to use various equation solving techniques (for example, standard unification, AC-unification, E-unification), to built-in specific equality handling rules (e.g., paramodulation), and to apply rewriting techniques and various tools of computer algebra systems.

In view of the third point, it is possible to say that we have all the necessary to begin the integration of deductive procedures with computer algebra systems.

As proof search methods relying on the principles described above can operate within the signature of an input problem, so it is possible to generate sufficient assertions and derive consequences in the form, which is usual for man. This enables to construct flexible interface tools.

Nowadays, there exists a series of projects and systems which have intersections with the EA programme in one way or another and which are close to it in spirit, for example, MIZAR, THEOREMA, OMEGA, ISABELLE, QED. A number of the above and some other projects, systems, and groups (for example, the DReaM group, Mechanized Reasoning Group, CAAR group, etc.) are interested in the integration of the deductive and computational power of both deduction systems and computer algebra systems. Taking this fact into consideration, the authors hope that this paper and some theses on the EA programme can be helpful in attacking such problems as distributed automated theorem proving, checking self-contained mathematical texts for correctness, remote training in mathematical disciplines, extracting knowledge from mathematical papers, and constructing data bases for mathematical theories.

References

[1] V. Glushkov. Foundations of mathematics and automation of deductive constructing (in Russian). In *Kibernetika. Voprosy teoriyi i praktiki*. Nauka, Moskva, 1986.

[2] A. Lyaletski and M. Morokhovets. On linguistic aspects of integration of computer mathematical knowledge. In *Proc. of CALCULEMUS Workshop*, Trento, Italy (1999) 165–178.

[3] V. Glushkov, K. Vershinin, et al. On a formal language for description of mathematical texts (in Russian). In *Avtomatizatsiya poiska dokazatel'stv teorem v matematike*. Institute of Cybernetics, Kiev (1974) 3–36.

[4] A. Degtyarev and A. Lyaletski. Logical inference in SAD (in Russian). In *Matematicheskiye osnovy sistem iskusstvennogo intellekta, Institute of Cybernetics*, Kiev, 1981.

[5] A. Degtyarev, A. Lyaletski, and M. Morokhovets. Evidence Algorithm and Sequent Logical Inference Search. In *LNAI* **1705** (1999) 44–61.

[6] F. Anufriyev, V. Fediurko, A. Letichevski, Z. Asel'derov, and I. Didukh. On one algorithm of theorem proof search in Group Theory (in Russian). *Kibernetika*, **1**, 1966, pp. 23 - 29.

[7] F. Anufriyev. An algorithm of theorem proof search in logical calculi (in Russian). In *Teoriya avtomatov*, **5**, Institute of Cybernetics, Kiev, 1969.

[8] V. Glushkov, V. Kostyrko, A. Letichevski, F. Anufriyev, and Z. Asel'derov. On a language for description of formal theories (in Russian). In *Teoreticheskaya kibernetika*, **3**, 1970.

[9] G. Gentzen. Untersuchungen über das logische Schliessen. *Math. Z.* **39** (1934) 176–210.

[10] T. Skolem. Logisch-kombinatorische Untersuchungen über die Erfüllbarkeit oder Beweisbarkeit mathematischer Sätze. *Skriftner utgit ar Videnskapsselskaper i Kristiania*, **4** (1920) 4–36.

[11] J. Herbrand. Recherches sur la Théorie de la Démonstration. *Travaux de la Societé des Sciences et de Lettres de Varsovie*, Class III, Sciences Mathématiques et Physiques (1930) Vol. 33.

[12] M. Davis and H. Putnam. A computing procedure for quantification theory. In *J. of the ACM* **7** (1960) 201–215.

[13] P. S. Gilmore. A proof method for quantification theory; its justification and realization. In *I.B.M.J. Res. Devl.* **4** (1960) 28–35.

[14] D. Prawitz. An improved proof procedure. In *Theoria* **26** (1960) 102–139.

[15] Wang Hao Towards mechanical mathematics. In *I.B.M.J. Res. Devl.* **4** (1960) 2–22.

[16] S. Kanger. Simplified proof method for elementary logic. In *Comp. Program. and Form. Sys.: Stud. in Logic. North-Holl.*, Publ. Co., 1963.

[17] J. Robinson. A machine-oriented logic based on resolution principle. In *J. of the ACM* (1965) 23–41.

[18] S. Maslov. An inverse method for deducibility checking in classical predicate calculus. In *Doklady Akademiyi Nauk SSSR* **159** (1964).

[19] A. Lyaletski. Gentzen calculi and admissible substitutions. In *Actes preliminaries, du Simposium Franco-Sovetique "Informatika-91"*. Grenoble, France (1991) 99–111.

[20] V. Atayan and M. Morokhovets. Combining formal derivation search procedures and natural theorem proving techniques in an automated theorem proving system. In *Cybernetics and System Analysts* **32** (1996) 442–465.

[21] A. Degtyarev, Yu. Kapitonova, A. Letichevsky, A. Lyaletski, and M. Morokhovets. A brief historical sketch on Kiev school of automated theorem proving. In *Proc. of the 2nd International THEOREMA Workshop*, Linz, Austria (June 1998) 151–156.

[22] J. Gallier. *Logic for computer science: foundations of Automatic Theorem Proving*. Harper and Row, Inc. New York (1986) 513 p.

[23] V. Atayan and K. Vershinin. On formalization of some inference search methods (in Russian). In *Avtomatizatsiya obrabotki matematicheskikh tekstov*. Institute of Cybernetics, Kiev (1980) 36–52.

Towards Learning New Methods
in Proof Planning

Mateja Jamnik   Manfred Kerber
Christoph Benzmüller

Abstract. *In this paper we propose how proof planning systems can be extended by an automated learning capability. The idea is that a proof planner would be capable of learning new proof methods from well chosen examples of proofs which use a similar reasoning strategy to prove related theorems, and this strategy could be characterised as a proof method. We propose a representation framework for methods, and a machine learning technique which can learn methods using this representation framework. The technique can be applied (amongst other) to learn whether and when to call external systems such as theorem provers or computer algebra systems.*

1   Introduction

Proof planning [2] is an approach to theorem proving which uses proof methods rather than low-level logical inference rules to prove a theorem at hand. A proof method specifies and encodes a general reasoning strategy that can be used in a proof, and hence represents a number of individual inference rules. For example, an induction strategy can be encoded as a proof method. Proof planners search for a proof plan of a theorem which consists of applications of several methods. An object-level logical proof can be generated from a proof plan. Proof planning is a powerful technique because it often dramatically reduces the search space, allows reuse of proof methods, and moreover generates proofs where the reasoning strategies of proofs are transparent, so they may have an intuitive appeal to a human mathematician.

One of the ways to extend the power of a proof planning system is to enlarge the set of available proof methods. Often a number of theorems can be proved in a similar way, hence a new proof method can encapsulate the general structure, i.e., the reasoning strategy of a proof for such theorems. A difficulty in applying a strategy to many domains is that in the current proof planning systems new methods have to be implemented and added by the developer of a system. Our aim is to explore how a system can learn new methods automatically given a number of well chosen examples of related proofs of theorems. This would be a significant improvement, since examples exist typically in abundance, while the extraction of methods from these examples can be considered as a major bottleneck of the proof planning methodology. In this paper we therefore propose an approach to automatic learning of proof methods within a proof planning framework.

There is a twofold, albeit loose, relation to the *Calculemus idea*. The first relation is with respect to Kowalski's equation *algorithm = logic + control* [9] – our work aims at exploring general reasoning patterns and the control knowledge hidden in the sets of well chosen proof examples. The extracted knowledge, which is in Kowalski's sense a form of algorithmic knowledge, is represented in such a way that it can be reused for tackling new problems. The exploration of algorithmic knowledge is especially desirable in cases when this knowledge is not a priori available to a reasoning system (e.g., in form of a built-in or connected computer algebra system). The second relation is that our approach, which is described and applied here solely in the context of simplification proof examples in group theory, is not restricted only to this domain. It can analogously be applied to learn control knowledge from proof examples that already contain calls to computer algebra systems. Therefore, note that computer algebra systems may be employed in a subordinated way in proof planning by calling them within proof methods to perform particular computations [8]. Given a set of examples each of which contains probably several such calls to computer algebra systems, our approach would enable a system to learn the overall pattern of these calls (if there is one). In this sense the learnt methods also encode knowledge of the controlled usage of computer algebra systems. More generally, this argument also applies to other external reasoning system which are subordinately employed within proof methods, and is not restricted to computer algebra systems only.

Figure 1 gives a structure of our approach to learning proof methods, and hence an outline of the rest of this paper. First, we give some background and motivation for our work. In Section 2 we examine what needs to be learnt, choose our problem domain and give some examples of proofs that use a similar reasoning strategy. Then, in Section 3, the representation of methods (in the form of method specifiers and method outlines) that renders the learning process as easy as possible is discussed. We continue in Section 4 to present one possible learning algorithm for learning proof methods from examples of proofs. Some alternative learning techniques are also discussed. Next, in Section 5, we revisit our method representa-

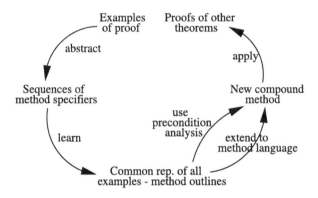

Figure 1. An approach to learning proof methods.

tion and enrich it so that the newly learnt methods can be used in a proof planner for proofs of other theorems. We use precondition analysis to acquire the information for extending the method representation. Finally, in Section 6, we relate our work to that of others, and conclude with some future directions and final remarks.

2   Motivating Example

A proof method in proof planning basically consists of a triple – precondition, postcondition and a tactic. A tactic is a program which given that the preconditions are satisfied executes a number of inference steps in order to transform an expression representing a subgoal so that the postconditions are satisfied by the transformed subgoal. If no appropriate method is available in a given planning state, then the user (in case of interactive systems) or the planner (in case of automated systems) has to explicitly apply a number of lower-level methods (with inference rules as the lowest-level methods) in order to prove a given theorem. It often happens that such a pattern of lower-level methods is applied time and time again in proofs of different problems. In this case it is sensible and useful to encapsulate this inference pattern in a new proof method. Such a higher-level proof method based on lower-level methods can either be implemented and added to the system by the user or the developer of the system. Alternatively, we propose that these methods could be learnt by the system automatically.

The idea is that the system starts with learning simple proof methods. As the database of available proof methods grows, the system can learn more complex proof methods. Initially, the user constructs simple proofs which consist only of basic inference rules rather than proof methods. A learning mechanism built into the proof planner then spots the proof pattern occurring in a number of proofs and extracts it in a new proof method. Hence, there is a hierarchy of proof steps from inference rules to complex methods. Inference rules can be treated as methods by assigning to them pre- and postconditions. Thus, from the learning perspective we can have a unified view of inference rules and methods as given sequences of elements from which the system is learning a pattern.[1]

To demonstrate our ideas with an example we need to first determine our problem domain – we choose theorems of abstract algebra, and in particular theorems of group theory. An example of a proof method is a simplification method which simplifies an expression using a number of simplification inference rules (which are in our unified view just basic-level proof methods).[2] In the case of group theorems, the simplification method may consist of applying both (left and right) axioms of identity, both axioms of inverse and the axioms of associativity. Note that e is the identity element, i is the inverse function, and LHS $\Rightarrow$ RHS stands for rewriting LHS to RHS:

$$
\begin{aligned}
(X \circ Y) \circ Z & \Rightarrow X \circ (Y \circ Z) & (20) \\
X \circ (Y \circ Z) & \Rightarrow (X \circ Y) \circ Z & (21) \\
e \circ X & \Rightarrow X & (22) \\
X \circ e & \Rightarrow X & (23) \\
X \circ X^i & \Rightarrow e & (24) \\
X^i \circ X & \Rightarrow e & (25)
\end{aligned}
$$

We now give two examples of proof steps which simplify given expressions and which are concrete applications of a simplification method that we want a system to learn.

---

[1]Note that as a consequence of the hierarchic character of the method language – with methods corresponding to calculus-level rules at the lowest-level – the approach is in principle general enough to learn methods on every level of abstraction. While some heuristic information for the compound method can be computed from the component methods, learning more precise heuristic information for the compound method will be necessary. We do not address this problem in this paper.

[2]One may assume that the simplification inference rules are already learnt as basic proof methods from rewriting style of proofs employing a single rewriting rule and appropriately instantiated group axioms.

Example 1

$$a \circ ((a^i \circ c) \circ b)$$
$$\Downarrow \ (\text{A-l})$$
$$(a \circ (a^i \circ c)) \circ b$$
$$\Downarrow \ (\text{A-l})$$
$$((a \circ a^i) \circ c) \circ b$$
$$\Downarrow \ (\text{Inv-r})$$
$$(e \circ c) \circ b$$
$$\Downarrow \ (\text{Id-l})$$
$$c \circ b$$

Example 2

$$a \circ (a^i \circ b)$$
$$\Downarrow \ (\text{A-l})$$
$$(a \circ a^i) \circ b$$
$$\Downarrow \ (\text{Inv-r})$$
$$e \circ b$$
$$\Downarrow \ (\text{Id-l})$$
$$b$$

In pseudocode, one application of simplification could be described as follows (notice that a repeated application of simplification can be learnt separately):

Precondition:
> *there are subterms in the initial term that are inverses of each other, and that are not separated by other subterms, but only by brackets.*

Tactic:
1. *apply associativity (A-r) and/or (A-l) for as many times as necessary (including 0 times) to bring the subterms which are inverses of each other together, and then*
2. *apply inverse inference rule (Inv-r) or (Inv-l) to reduce the expression, and then*
3. *apply the identity inference rule (Id-r) or (Id-l).*

Postcondition:
> *the initial term is reduced, i.e., it consists of fewer subterms.*

Note that this is a general simplification method, and the two examples given above would not be sufficient to learn it (e.g., the examples do not use (A-r), (Inv-l), (Id-r), hence additional examples that use these inference rules would have to be provided). Furthermore, our simplification method needs to be able to do loop applications of methods, where the number of loops is determined by the theorem the method is used to prove. In a sense, this type of control construct is similar to the notion of tacticals, hence we refer to them as methodicals. We are realising our ideas on learning methods in the proof planner of ΩMEGA [1] which does not explicitly represent loops. Loops in ΩMEGA are simulated by the use of control rules (see [4]).

Therefore, we are currently extending the existing representations used in ΩMEGA to provide explicit representation of loop applications of methods.

An alternative to this approach is to re-represent our problems, for instance, so as to omit the brackets in the presence of the associativity rules. However, this would be a much harder learning problem. In this paper we do not take this approach.

3 Method Outline Representation

The representation of a problem is of crucial importance for solving it – a good representation of a problem renders the search for its solution easy. This is a well known piece of advice from Pólya [14]. The difficulty is in finding a good representation. Our problem is to devise a mechanism for learning methods. Hence, the representation of a method is important and should make the learning process as easy as possible. Furthermore, it should be possible to represent loop applications of inference rules in methods. Here we present a simple representation formalism for methods, which abstracts away as much information as possible for the learning process, and then restores the necessary information so that the proof planner can use the newly learnt method. At the same time it caters for loop applications of inference rules in a method.

A major problem we are faced with when we want to learn compound methods from lower-level ones is the intrinsic complexity of methods, which goes beyond the complexity that can typically be tackled in the field of machine learning. For this reason we first simplify the problem by trying to learn the so-called *method outlines* (which is discussed next), and second, we reconstruct the full information by extending outlines to methods using precondition analysis (which will be discussed in Section 5).

Let us assume the following language L, where P is a set of primitives (which are the known identifiers of methods used in a method that is being learnt). In essence, this language defines regular expressions over method identifiers. The weight w defines the complexity of an expression:

- for any $p \in P$, let $p \in L$ and $w(p) = 0$,
- for any $l_1, l_2 \in L$, let $[l_1, l_2] \in L$ and $w([l_1, l_2]) = w(l_1) + w(l_2) + 1$,
- for any $l_1, l_2 \in L$, let $[l_1|l_2] \in L$ and $w([l_1|l_2]) = w(l_1) + w(l_2) + 1$,
- for any $l \in L$, let l^* and $w(l^*) = w(l) + 1$.

"[" and "]" are auxiliary symbols used to separate subexpressions, "|" denotes an *exclusive-or-disjunction*, "," denotes a *sequence*, and "*" denotes a *repetition* of a subexpression any number of times (including 0). Let the set of primitives P be $\{\text{A-l}, \text{A-r}, \text{Inv-l}, \text{Inv-r}, \text{Id-l}, \text{Id-r}\}$. Using this language, and given the appropriate pre- and postconditions, the tactic of our simplification method described above could be expressed as:

$$simplify \equiv \Big[\big[[\text{A-l} \mid \text{A-r}]^*, \; [\text{Inv-l} \mid \text{Inv-r}]\big], \; [\text{Id-l} \mid \text{Id-r}] \Big]$$

with $w(simplify) = 6$. We refer to expressions in language L which describe compound methods (as in the example above) as *method outlines*. *simplify* is a typical method outline that we would like our system to be able to learn automatically. The representation is simple enough that a mechanism can be devised to learn methods using this representation. We propose such a mechanism next.

4 How to Learn?

As explained in the previous section, we use our language L for an abstract representation of methods, i.e., method outlines, in order to simplify the representation and render the learning process easier. Here we address how such method outlines can be learnt given a number of well chosen examples of proofs. Typically, there are many possible method outlines which describe a method that the system is learning. One possible approach is to learn the simplest (following Occam's razor principle) and optimal method outline, and we measure simplicity (in our first approach) in terms of weight as defined before in Section 3. Notice that our decision to use simplicity to select a possible method outline is a heuristic choice which in some cases may be inappropriate. That is, sometimes we do not prefer the simplest method outline, but perhaps the most general one, or even the least general one. As part of our future work we may have to refine our notion of simplicity.

We identified three possible approaches to learning [7]: a complete generation of method outlines, a guided generation of method outlines, and a technique similar to least general generalisation.

Exhaustive generation of all method outlines is our first attempt at a learning technique. The technique generates all possible method outlines in language L of a certain weight. We then prune this set by eliminating the method outlines which could not be instantiated to the representation of the examples. The size of the set of possible method outlines increases hyperexponentially which poses a severe problem. We can improve upon this performance by generating only the relevant method outlines – this is our second attempt at a learning technique. For instance, given any example, we could avoid generating all those method outlines that cannot be part of the method outline to be learnt (*e.g.*, if the sequence Inv-r, Inv-r does not occur in any of the training examples then no method outline containing this sequence should be generated). This would prune the possibilities dramatically, however, it should be further examined to see if this is enough.

Although these two learning mechanisms are rather inefficient, they form a standard which can be used to measure the quality of a solution found by a procedure which is computationally more efficient.

Generalisation A third possible approach for a learning technique is similar to Plotkin's least general generalisation [12, 13]. Let us assume a few good examples of method outlines and the sequence of inference rules that are applied on them in order to simplify them:

1. $a \circ ((a^i \circ c) \circ b)$ which is simplified by the following sequence of application of rules [A-l, A-l, Inv-r, Id-l].

2. $a \circ (a^i \circ c)$ which is simplified by the application of [A-l, Inv-r, Id-l]. Using a sort of heuristically guided generalisation similar to least general generalisation we would get the following method outline from the first two examples: $\big[$A-l*, Inv-r, Id-l$\big]$.

3. $a^i \circ (a \circ c)$ which is simplified by the application of [A-l, Inv-l, Id-l]. Adding this example, the method outline is generalised to $\big[$A-l*, [Inv-r|Inv-l], Id-l$\big]$.

4. $b \circ (a \circ a^i)$ which is simplified by the application of [Inv-r, Id-r]. The method outline is now generalised to $\big[$A-l*, [Inv-r|Inv-l], [Id-l|Id-r]$\big]$. Notice that the first generalisation of A-l to A-l* is still okay here.

5. $(b \circ (c \circ a)) \circ a^i$ which is simplified by the application of [A-r, A-r, Inv-r, Id-r]. The method outline is now generalised to $\big[$[A-l*|[A-r, A-r]], [Inv-r|Inv-l], [Id-l|Id-r]$\big]$.

6. $(b \circ (c \circ (d \circ a))) \circ (a^i \circ f)$ which is simplified by the application of [A-r, A-r, A-r, A-l, Inv-r, Id-r]. The method outline is finally generalised to $\big[$[A-l|A-r]*, [Inv-r|Inv-l], [Id-l|Id-r]$\big]$.

Such generalisation is not guaranteed to produce an optimal solution or even any solution at all, however it *is* a computationally *feasible* technique. Devising an effective and efficient algorithm for heuristically guided generalisation will form part of our future work. Here are some preliminary ideas. There are two general principles that can be employed in the generalisation:

- whenever a method M is applied a varied number of times in different examples (e.g., A-r is sometimes applied once, sometimes it is applied twice or not at all), then the application of that method is generalised into a starred method outline M^*,

- whenever the methods used are different, then they are generalised into a disjunction of these methods (e.g., if in one example we use Id-l and in another at the same point in the proof we use Id-r, then these are generalised into $[\text{Id-l}|\text{Id-r}]$).

The difficult part is to distinguish between the different stages of the application of methods in the proof. For instance, using the generalisation of the first two examples above, $[\text{A-l}^*, \text{Inv-r}, \text{Id-l}]$, and the third example, $[\text{A-l}, \text{Inv-l}, \text{Id-l}]$, there are more possible generalisations than just the following: $[\text{A-l}^*, [\text{Inv-r}|\text{Inv-l}], \text{Id-l}]$. For instance, this is also possible: $[\text{A-l}^*, \text{Inv-r}^*, \text{Inv-l}^*, \text{Id-l}]$. The reason that the first generalisation was chosen is that we know that the method can be split into three parts: the first applies associativity rules, the second applies the inverse rules and the third applies the identity rules.[3] But how can we find such a partition of a method that we want to learn in general and automatically? The precondition analysis [15, 5] may help us here – we discuss this next.

5 Method Representation Revisited

Methods expressed using the language introduced in Section 3 do not specify what their preconditions and postconditions are. They also do not specify how the number of loop applications of inference rules is instantiated when used to prove a theorem. Hence, the method representation needs to be enriched to account for these factors. We propose to use the ideas from precondition analysis developed by Silver [15] and later extended by Desimone [5] in order to enrich our method representation.

5.1 Precondition Analysis

The idea of precondition analysis is to examine the reasons for applying each inference at each step of the proof. This is achieved by providing explanations for each step in the proof which are usually extracted from the information of preconditions and postconditions of a step. The preconditions of each rule used in a method are paired with additional information, namely the methods that generated these preconditions. Similarly, the postconditions of each rule used in a method are paired with the methods that use these postconditions. We extend Desimone's *method schema* representation with the *effects* that an inference rule has in the proof. Effects are used to express a change in the proof planning state which is not explicitly planned for, because, for instance, the underlying language may not be

---

[3]Notice also that the chosen method outline is of smaller weight. However, weight as defined before, (i.e., complexity) may not always be the best heuristic choice.

rich enough to express these changes. We demonstrate a representation of a method schema with an example.

Let there be a proof of a theorem T which consists of three steps, M_1, M_2 and M_3. These methods consist of preconditions, postconditions, effects and tactics as demonstrated in the table below:

	Preconditions	Postconditions	Effects	Tactic
M_1	$x \wedge y \wedge p(d)$	$x \wedge y \wedge z$	$w = f(d)$	t_1
M_2	$y \wedge p(w)$	m		t_2
M_3	$z \wedge m$	n		t_3

Notice that $p(d)$ denotes some property p of d, where this property is a precondition of M_1. f denotes a function which under the application of M_1 changes a term d occurring in a precondition $p(d)$ to a term w. The initial state of the proof of the theorem can be described in terms of $x \wedge y \wedge p(d)$ which holds for theorem T. The preconditions of each method M_i used in a proof of an example are analysed to determine the explanations for using these particular steps in the proof. The explanations are generated in a bottom-up fashion starting with the application of the final method. M_3 was obviously applied in order to reach a solution denoted by n and is not an interesting case. Now consider reasons for applying method M_2. This is done by the analysis of the preconditions of M_3 which may be provided as the postconditions of M_2. The preconditions for method M_3 are $z \wedge m$. The precondition z was already satisfied before the application of M_2,[4] but the precondition m was generated as a postcondition of the application of method M_2. Hence, one possible explanation is that method M_2 was applied in order to generate precondition m for method M_3. The same can be said for method M_1 – it is applied in order to generate an effect w whose property $p(w)$ is a precondition for M_2, and a precondition z for M_3. Together, an explanation can be that methods M_1 and M_2 are applied in order to generate the preconditions $z \wedge m$ for M_3.[5] Desimone captures these explanations in a method schema.

---

[4]Note that all changes to the state of the proof caused by the application of a method have to be explicitly stated in the specification of the method (i.e., in its pre-, postconditions or effects). For example, nothing is mentioned about z in the specification of M_2, hence the application of M_2 preserves z.

[5]Using the precondition analysis as explained in [15], we get the explanations for applying all of the inference rules used in the examples. In our example, the associativity inference rules are applied to generate the preconditions for the inverse inference rules. Furthermore, the inverse inference rules are used to generate the preconditions for the identity rules. Hence, this gives the partition of the rules that we need in the generalisation learning process, as discussed in Section 4. The details of how the partitioning is done still need to be worked out.

Pre	$x \wedge y$		
		Preconditions	Postconditions
Tactic	M_1	$pre(x, \_), pre(y, \_)$ $pre(p(d), \_)$	$post(x, \_), post(y, M_2),$ $post(z, M_3)$ Effects: $w = f(d)$
	M_2	$pre(y, M_1), pre(p(w), M_1)$	$post(m, M_3)$
	M_3	$pre(z, M_1), pre(m, M_2)$	$post(n, \_)$
Post	n		

Figure 2. Compound method.

Using the language for method outlines described in Section 3 we are able to re-represent methods in the way suggested by the precondition analysis. In our example above, the method schema, say M_4 which is $[M_1, M_2, M_3]$ in our language L, can be re-represented as a method as given in Figure 2, where "*pre*" is a predicate with two arguments: the precondition and the method which created this precondition as a postcondition; "*post*" is a predicate with two arguments: the postcondition and the method which uses this postcondition as a precondition; "_" stands for no method, that is, the precondition is true before the application of the new method or the postcondition is not used as a precondition for any other method used in the tactic.

Analogously, a method outline with a disjunction such as $[M_1|M_2]$ can be represented as a method schema, the precondition of which is a disjunction of preconditions for M_1 and M_2. Similarly, the postcondition of this method schema is a disjunction of postconditions of M_1 and M_2. The planner has to be able to handle disjunctions of pre- and postconditions, which is a non-trivial open problem in proof planning. Extending a planner to deal with such disjunctions needs to be addressed in detail in the future. The second argument of pre- and postcondition pairs is determined as explained above. We further extend Desimone's method schema representation with a disjunction of explanations for method applications. Namely, if there is more than one method which generates/uses a pre-/postcondition, then these are combined disjunctively. This allows us to encode the fact that a postcondition of a particular rule is also a precondition of the same rule, hence, the rule can be applied several times.

Depending on the example that the method is used to prove, the pre- and postcondition pairs are instantiated differently. This determines if and how many repeated applications of a rule are needed. Hence, a method outline with a "*" such as $[M_1^*, M_2]$ has $pre(x, \_ \vee M_1), pre(y, \_ \vee M_1)$ as preconditions of M_1, and $post(x, \_ \vee M_1), post(y, M_1 \vee M_2), post(z, \_)$ as postconditions of M_1 (the rest is as expected). Hence, if after the application of M_1 all the postconditions of M_1 satisfy its preconditions, and furthermore no other method in the tactic is applicable then M_1 is applied again. If its postconditions satisfy the preconditions of another method, say M_2, then M_1 is no longer applied in the proof. This is of course a heuristic decision and further research will have to examine whether it is appropriate.

5.2 From Outlines to Methods

Now we can represent a method outline

$$\Big[[\text{A-l}|\text{A-r}]^*, [\text{Inv-r}|\text{Inv-l}], [\text{Id-l}|\text{Id-r}] \Big]$$

as a simplification method using pairs of pre- and postconditions with methods that generate or use them, and the effects that the rules have. To save space, but still convey the main points, we only consider a simplified version of simplification, namely a method outline $\big[\text{A-l}^*, [\text{Inv-r}|\text{Inv-l}], \text{Id-l}\big]$. Notice that in order to be able to attach explanations to the inference rules in the style of precondition analysis, the method language needs to be extended.

We extend it with the following vocabulary: $subt(X, Y)$ for "X is a subterm of Y", $nb(X, Y)$ for "X is a neighbour of Y" (two subterms of a term are neighbours if they are listed one after another when a tree representing the term is traversed in post-order), $\delta(X, Y, Z)$ for "a distance between subterms X and Y in a term Z" (the distance between two subterms is the number of nodes between the two subterms when a tree representing a term is traversed in post-order decreased by 1), and $red(E_1, E_2)$ for the fact that an expression E_1 is reduced to E_2 (an expression is reduced when it consists of fewer subterms than originally).[6] Figure 3 shows the relevant inference rules augmented with appropriate explanations. E' is the term generated from E by applying a corresponding method.

Figure 4 gives a method schema representation for a method outline $\big[\text{A-l}^*, [\text{Inv-r}|\text{Inv-l}], \text{Id-l}\big]$.[7] Note that $applicable(x)$ means that all the preconditions for x are satisfied. Additional information that all methods assume is a position parameter which specifies a subterm on which a method is applied. This information is used in the expansion of the method to the lower-layer methods. That is, should the user want an object-level proof from a proof plan, then ultimately all the methods need to be expanded to the inference rule level. Therefore, our new *simplify* method also requires

---

[6]The choice of this vocabulary is not important for this paper, and needs to be discussed elsewhere.

[7]Notice in Figure 4 that in the application of (A-l), A_1 matches with A and A_2 matches with A^i, or A_1 matches with A^i and A_2 matches with A.

Rule	Preconditions	Postconditions
(A-r)	$subt((X \circ Y) \circ Z, E) \wedge$ $subt(A, Y) \wedge subt(B, Z) \wedge$ $nb(A, B, E) \wedge \delta(A, B, E) > 0$	$subt(X \circ (Y \circ Z), E') \wedge$ $subt(A, Y) \wedge$ $subt(B, Z) \wedge nb(A, B, E')$ Effects: $\delta(A, B, E') =$ $\delta(A, B, E) - 1$
(Id-l)	$subt(e, E)$	$red(E, E')$
(Inv-r)	$subt(X \circ X^i, E) \wedge subt(X, E) \wedge$ $subt(X^i, E) \wedge$ $nb(X, X^i) \wedge \delta(X, X^i, E) = 0$	$subt(e, E')$ $red(E, E')$
(Inv-l)	$subt(X^i \circ X, E) \wedge subt(X, E) \wedge$ $subt(X^i, E) \wedge$ $nb(X, X^i) \wedge \delta(X, X^i, E) = 0$	$subt(e, E')$ $red(E, E')$

Figure 3. Methods with explanations for their application.

the position parameter information. Details about the extraction of a position parameter still need to be resolved, hence we do not discuss them here.

<table>
<tr><td>Pre</td><td colspan="3">$nb(A, A^i, E) \wedge (applicable(\text{A-l}) \vee applicable(\text{Inv-r}) \vee applicable(\text{Inv-l}))$</td></tr>
<tr><td rowspan="5">Tactic</td><td>Tact</td><td>Preconditions</td><td>Postconditions</td></tr>
<tr><td>(A-l)</td><td>$pre(subt(K \circ (L \circ M), E_1),\ _ \vee (\text{A-l}))$
$pre(subt(A_1, K),\ _ \vee (\text{A-l}))$
$pre(subt(A_2, L),\ _ \vee (\text{A-l}))$
$pre(nb(A, A^i, E_1),\ _ \vee (\text{A-l}))$
$pre(\delta(A, A^i, E_1) > 0,\ _ \vee (\text{A-l}))$</td><td>$post(subt((K \circ L) \circ M, E_2), _)$
$post(subt(A_1, K),$
$(\text{A-l}) \vee (\text{Inv-r}) \vee (\text{Inv-l}))$
$post(subt(A_2, L),$
$(\text{A-l}) \vee (\text{Inv-r}) \vee (\text{Inv-l}))$
$post(nb(A, A^i, E_2),$
$(\text{A-l}) \vee (\text{Inv-r}) \vee (\text{Inv-l}))$
Effects: $\delta(A, A^i, E_2) =$
$\delta(A, A^i, E_1) - 1$</td></tr>
<tr><td>(Inv-r)</td><td>$pre(subt(A \circ A^i, E_3),\ _ \vee (\text{A-l}))$
$pre(subt(A, E_3),\ _ \vee (\text{A-l}))$
$pre(subt(A^i, E_3),\ _ \vee (\text{A-l}))$
$pre(nb(A, A^i),\ _ \vee (\text{A-l}))$
$pre(\delta(A, A^i, E_3) = 0,\ _ \vee (\text{A-l}))$</td><td>$post(subt(e, E_4),\ (\text{Id-l}))$
$post(red(E_3, E_4),\ _)$</td></tr>
<tr><td>(Inv-l)</td><td>$pre(subt(A^i \circ A, E_5),\ _ \vee (\text{A-l}))$
$pre(subt(A, E_5),\ _ \vee (\text{A-l}))$
$pre(subt(A^i, E_5),\ _ \vee (\text{A-l}))$
$pre(nb(A, A^i),\ _ \vee (\text{A-l}))$
$pre(\delta(A, A^i, E_5) = 0,\ _ \vee (\text{A-l}))$</td><td>$post(subt(e, E_6),\ (\text{Id-l}))$
$post(red(E_5, E_6),\ _)$</td></tr>
<tr><td>(Id-l)</td><td>$pre(subt(e, E_7),\ (\text{Inv-r}) \vee (\text{Inv-l}))$</td><td>$post(red(E_7, E_8), _)$</td></tr>
<tr><td>Post</td><td colspan="3">$red(E, E_i)$</td></tr>
</table>

Figure 4. Newly learnt compound method *simplify*.

6 Further Work and Conclusion

In this paper we introduced a language for method outlines which can be used for describing compound proof methods in proof planning on an abstract level. These methods can carry out loop applications of less complex methods and can apply them disjunctively, depending on the theorem for which they are used to prove. We also introduced a technique for learning method outlines from a number of examples of method applications. The heuristics for this generalisation technique need to be worked out in the future.

The method outlines of the introduced language can be encoded as proof method schemas in the style of Silver and Desimone in their work on precondition analysis. We demonstrated how a method outline learnt from a number of examples of the simplification method can be represented as a method schema.

Our approach is restricted to learning new higher-level proof methods on the basis of the already given ones. We cannot learn language extensions such as a coloured term language which would be a prerequisite for learning any kind of methods similar to rippling [3]. In this paper we do not address the question how such a vocabulary can be learnt by a machine. Work by Furse on MU learner [6] may be relevant for this task.

Not much work has been done in the past on applying machine learning techniques to theorem proving, in particular proof planning. We already mentioned work by Silver [15] and Desimone [5] who used precondition analysis to learn new method schemas – we explained how we use their ideas in our work. Of interest is work on generalisation [13], and other machine learning techniques such as inductive logic programming [11] and explanation based generalisation [10].

Finally, there are many open questions that remain to be answered. Here are some of them:

- Are the descriptions of preconditions, postconditions and effects as given in the example in this paper adequate to describe methods? Will a proof planner be able to use such method schemas in order to instantiate them into methods, and hence prove theorems? What type of extensions of a proof planner are needed to accommodate the use of method schemas?

- Does the representation of the method schema given in the example in this paper adequately describe the method outline represented using our language L? Can the method schema representation be simplified?

- So far we considered proofs which are constructed in a purely sequential rewriting style without any case splits; i.e., we considered proof chains and not general proof trees. Does our approach fully apply also to styles of proofs other than rewriting?

- Our recent experiments showed that our primitive proof methods are assumed to focus on particular subterm occurrences. These are specified by additional position parameters provided by the methods. Furthermore, the learnt *simplify* method has to come with an additional position parameter which indicates where in the expression of a theorem the method is applied. How can the parameters required for a new method be inferred from the parameters given for the primitive methods? How can the learnt method generally guide its expansion to an object-level proof by providing appropriate parameter information to the primitive methods?

- How can we most efficiently learn a general method outline in language L describing a method schema? Which one of the two proposed approaches, namely guided generation of all method outlines describing the example and then pruning them, and a technique similar to Plotkin's least general generalisation, is best to use? Are there examples for which the first technique is better than the second, and vice versa? How can we determine when a technique is or is not appropriate? Is there another, more appropriate technique that we could use in order to learn new methods automatically?

Some of the answers to these questions have the potential to significantly contribute to the strength of the proof planning approach to mechanised reasoning.

Acknowledgements

We would like to thank Alan Bundy for his continuing interest in and advice on our work, and in particular for pointing us to the work of Silver and Desimone. Furthermore we would like to thank Andreas Meier and Volker Sorge for their invaluable help in starting to implement our ideas in ΩMEGA. This work was supported by EPSRC grants GR/M22031 and GR/M99644.

References

[1] C. Benzmüller, L. Cheikhrouhou, D. Fehrer, A. Fiedler, X. Huang, M. Kerber, M. Kohlhase, E. Melis, A. Meier, W. Schaarschmidt,

J. Siekmann, and V. Sorge. Ω: Towards a mathematical assistant. In W. McCune, editor, *14th Conference on Automated Deduction*, pp. 252–255, 1997.

[2] A. Bundy. The use of explicit plans to guide inductive proofs. In R. Lusk and R. Overbeek, editors, *9th Conference on Automated Deduction*, pages 111–120. Springer Verlag, 1988. Longer version available from Edinburgh as DAI Research Paper No. 349.

[3] A. Bundy, A. Stevens, F. van Harmelen, A. Ireland, and A. Smaill. Rippling: A heuristic for guiding inductive proofs. *Artificial Intelligence*, 62:185–253, 1993. Also available from Edinburgh as DAI Research Paper No. 567.

[4] L. Cheikhrouhou. *Forthcoming*. PhD thesis, Fachbereich Informatik, Universität des Saarlandes, Saarbrücken, Germany, 2000.

[5] R.V. Desimone. Learning control knowledge within an explanation-based learning framework. In I. Bratko and N. Lavrač, editors, *Progress in Machine Learning – Proceedings of 2nd European Working Session on Learning, EWSL-87*, Bled, Yugoslavia, 1987. Sigma Press. Also available from Edinburgh as DAI Research Paper 321.

[6] E. Furse. The mathematics understander. In J.H. Johnson, S. McKee, and A. Vella, editors, *Artificial Intelligence in Mathematics*. Clarendon Press, Oxford, 1994.

[7] M. Jamnik, M. Kerber, and C. Benzmüller. Towards Learning New Methods in Proof Planning. Technical Report, School of Computer Science, The University of Birmingham, CSRP-00-9, 2000.

[8] M. Kerber, M. Kohlhase, and V. Sorge. Integrating Computer Algebra Into Proof Planning. *Journal of Automated Reasoning*, 21(3):327–355, 1998.

[9] R. Kowalski. Algorithm = Logic + Control. *Communications of the Association for Computing Machinery*, 22:424–436, 1979.

[10] T.M. Mitchell, R.M. Keller, and S.T. Kedar-Cabelli. Explanation-based generalization: A unifying view. *Machine Learning*, 1(1):47–80, 1986. Also available as Tech. Report ML-TR-2, SUNJ Rutgers, 1985.

[11] S.H. Muggleton and L. De Raedt. Inductive logic programming: Theory and methods. *Journal of Logic Programming*, 19, 20:629–679, 1994.

[12] G. Plotkin. A note on inductive generalization. In D. Michie and B. Meltzer, editors, *Machine Intelligence 5*, pages 153–164. Edinburgh University Press, 1969.

[13] G. Plotkin. A further note on inductive generalization. In D. Michie and B. Meltzer, editors, *Machine Intelligence 6*, pages 101–126. Edinburgh University Press, 1971.

[14] G. Pólya. *How to solve it*. Princeton University Press, 1945.

[15] B. Silver. Precondition analysis: Learning control information. In R.S. Michalski, J.G. Carbonell, and T.M. Mitchell, editors, *Machine Learning 2*. Tioga Publishing Company, 1984.

Using Meta-variables for Natural Deduction in *Theorema*

Boris Konev*   Tudor Jebelean*

Abstract. *We approach the problem of finding witnessing terms in proofs by the method of meta-variables. We describe an efficient method for handling meta-variables in natural style proofs and its implementation in the* Theorema *system. The method is based on a special technique for finding meta-substitutions when the proof search is performed in an **AND-OR** deduction tree. The implementation does not depend on the search strategy and allows easy integration with various special provers as well as with special unification/solving engines.*

1 Introduction

The problem of generating proofs that can be easily read by non-specialists has raised an increased interest in the recent years (see e.g. [6, 12, 5, 19]). One of the main advantages of "natural style" proving is that one can implement as proof procedures the techniques used by mathematicians, and one can also use advanced solvers for particular domains, in the form of special inference rules. However, the flexibility needed for dynamic extension of the prover with special inference rules, as well as for dynamic interaction with special "black box" computing and solving engines makes it difficult to use established efficient techniques such as resolution, tableaux, or connection methods (see e.g. [6] for a survey). Therefore the usual approach is to employ a sequent calculus (like Gentzen or similar) — whose main draw-back is the lack of determinism in selecting the right replacements in some of the quantified rules (elimination of ∀ in assumptions and of ∃

---

*Supported by INTAS, Project 96-0760 and Austrian Science Foundation (FWF), SFB project F1302.

in goal). The corresponding problem in "natural style" deduction is the finding of "witnessing terms" (especially for existential goals).

In fact, human produced proofs often proceed by denoting such a desired witness by a variable at the meta-level, which is instantiated later in the proof with a suitable term suggested by the inferred knowledge. (When working in special domains the value may be even obtained by using a special solver—see for instance the concept of PCS proving [10] or [19].) In formal reasoning this has been formalized as the method of *meta-variables* (also called *parameters, dummy variables, free variables*—see [21, 22, 17, 8, 7, 15]).

In this paper we present an algorithm for handling meta-variables in natural style proofs and its implementation in the *Theorema* system. The proving style of *Theorema* [9, 12, 10] imitates the proving style of the working mathematician, and the system contains special provers and solvers for various domains, new elements being constantly added. Therefore, the system already uses a mechanism for combining various provers and solvers. We build upon this mechanism by adding the functionality which is needed for handling meta-variables, thus obtaining an implementation of the method which has the following distinctive characteristics:

- The method can be used together with any set of inference rules, in particular with any special provers.

- The method can be used together with any special (constraint) solvers and rewrite engines based on implicit or explicit equality sets.

- The proof is obtained as soon as possible, and inconsistent meta-substitutions are eliminated as soon as they can be detected.

- Proof sub-trees are not destroyed, but they are reused as much as possible.

- The method can be used together with any strategy for constructing the proof tree.

In any implementation of the meta-variables method, one constructs first a (partial) proof tree which contains meta-variables — this is usually called a *pre-proof*. A substitution of concrete terms instead of meta-variables maps the pre-proof into a concrete proof.

In free tableaux calculus (see e.g. [3]), the choice of substitution is done non-deterministically and the substitution is applied immediately to the whole pre-proof. The indeterminism needs to be resolved by *enumeration* of all possible substitutions and *backtracking*. Usually an AND-OR tree (of proofs) is used for this purpose (cf. [3]). The main disadvantage of such a method is that the information on inference steps may have been lost after backtracking—this is why this kind of methods are also called *destructive*. Recently, complete destructive procedures that avoid backtracking have appeared [4] (similar ideas can be found in [2]). The essence of these procedures is that whenever a substitution is applied, new subtrees are added to all affected branches of the deduction tree. These subtrees "reconstruct" the destroyed formulas. The search space is reduced at the cost of increasing the size of the deduction tree.

A straightforward way to enforce non-destructiveness for tableaux calculus is described in [14, 15]. The method consists of delaying any substitution until all branches can be closed simultaneously. This requires termination testing after each deduction step.

In this paper we develop an efficient non-destructive procedure with restricted backtracking. Instead of constructing an AND-OR tree on the top of possible deduction trees, the deduction tree of *Theorema* itself contains AND and OR nodes, they become a part of the calculus. As in [14, 15], we associate with each node of a pre-proof the substitution which makes the subtree a correct proof. Then, we "lift" substitutions bottom-up along the pre-proof. If eventually we succeed to lift the substitution up to the root, the proof search is finished. Note, that a subtree of a pre-proof can be mapped to a correct proof with different substitutions, we keep all of them.

While substitutions are lifted, they are tested for compatibility. If at certain node the substitutions assigned to its successors are incompatible, the proof search should be continued somewhere under this node. In addition to substitutions, we pass bottom-up positions of "open" branches, we use this positions to find the place where proof search should be continued.

The *Theorema* system is a result of work of the *Theorema* group (see http://www.theorema.org) based on earlier work of Buchberger [9]. The control procedure is mainly written by Tomuta [13], the basic prover for Natural Deduction was developed by Buchberger and Jebelean [11]. The meta-variable method was incorporated into the system by the first author of this paper.

2   Meta-variables in Natural Deduction

We consider a Natural Deduction calculus operating on *proof situations*, which are represented as a *sequents* of the form

$$A_1, A_2, \ldots, A_n \vdash G \text{ or } \Gamma, A \vdash G$$

where $A_1, A_2, \ldots, A_n, A$ denote assumptions, Γ denotes several assumptions, and G is the goal (to be proven).

In order to construct the proof of a sequent, we recursively transform it to one or several sequents until we reduce it to a set of obviously valid sequents (*final sequents*). Reductions are performed by means of *inference rules* of the form

$$\frac{S_1; \quad S_2; \quad \ldots; \quad S_n}{S}$$

(sequent S reduces to sequents $S_1, \ldots, S_n$). Usually (and in this paper) one only considers rules where S is valid iff all S_i are valid: see Section 4. A proof is a tree of sequents whose root is the initial sequent and the leaves are final sequents.

The "difficult" rules that eliminate quantifiers (cf. [16]) are:

$$\frac{\Gamma \vdash [F]_{x \to t} \vee \exists x F}{\Gamma \vdash \exists x F}, \qquad \frac{\Gamma, [F]_{x \to t} \wedge \forall x F \vdash G}{\Gamma, \forall x F \vdash G}, \qquad (26)$$

where t is a term, x is an object variable, t is free for x in F, and $[F]_{x \to t}$ is the result of substituting the term t in place of all free occurrences of x in F. The existential goal is duplicated for the sake of completeness of the calculus, but for many problems it is enough to find a single witnessing term – this is controlled in our prover by an option.

The main idea of the meta-variable method is to postpone the construction of the concrete term t until the moment when the proof search suggests an appropriate one. This is done by replacing t in (26) by a new meta-variable ξ, whose value is to be found in the proof process:

$$\frac{\Gamma \vdash [F]_{x \to \xi} \vee \exists x F}{\Gamma \vdash \exists x F}, \qquad \frac{\Gamma, [F]_{x \to \xi} \wedge \forall x F \vdash G}{\Gamma, \forall x F \vdash G}. \qquad (27)$$

The proof-tree obtained using rules (2) is called a *pre-proof*, which is called *correct* if there exist a *correct* substitution for all meta-variables which transforms the pre-proof into a real proof. Therefore, a substitution is correct iff **(a)** all intermediate rule applications are correct; and **(b)** final nodes are valid.

Concerning (a), the only rules that may become invalid after applying a substitution are quantifier rules, namely (27) and

$$\frac{\Gamma \vdash [F]_{x \to a}}{\Gamma \vdash \forall x F}, \qquad \frac{\Gamma, [F]_{x \to a} \vdash G}{\Gamma, \exists x F \vdash G}. \qquad (28)$$

(Note that we use Skolem constants instead of functions, because this leads to more natural proofs.) Following e.g. [1], it is enough to require in (28) that $\forall x F$ (respectively, $\exists x F$) does not contain a. Suppose that $\forall x F$ ($\exists x F$)

contains a meta-variable ζ. A substitution is correct if its value on ζ does not contain a. We denote the described type of restrictions by

$$a \notin \zeta. \tag{29}$$

The rules (26) require the term t to be free for x in F. When we replace the rules (26) with (27), this restriction turns into restrictions similar to (29) (for details, see Section 4). The union of the sets of restrictions corresponding to each application of a quantifier rule forms a set of restrictions for the pre-proof.

Concerning (b), we use a calculus whose only final proof situation are $\Gamma, A \vdash A$, and $\Gamma, A, \neg A \vdash G$. Therefore, a leaf of a pre-proof is transformable into a final proof situation if and only if one of the assumptions is unifiable with the goal ("the goal is an assumption") or an assumption is unifiable with the negation of an assumption ("assumptions are contradictory"). The union of all such pairs of formulas forms a *unification problem* for the pre-proof.

It was independently proved by Orevkov [20], and Krajíček and Pudlák [18] that if a most general unifier for the unification problem exists and satisfies the set of restrictions, then it is a correct substitution.

3   The Proof Search Procedure of *Theorema*

The proof search procedure of *Theorema* is described in detail in [13]. The successive reductions (and finally the proof) are represented in as an AND-OR *deduction tree*. AND nodes correspond to the usual sequent reductions (see Section 2), while OR nodes represent proof alternatives.

Theorema can combine several proof engines called **basic provers**, some of them specialized for certain domains. Basic provers return subtrees representing one or several deduction steps. A **control procedure** activates the various basic provers, combines their output and thus constructs the proof-tree in a top-down inductive manner. The communication between provers is realized by means of a global and a local *proof context*, containing various information associated to the whole proof or to the particular branch, respectively. A *proof value* is assigned to each node of a deduction tree. The proof value of a node can be *proved*, *failed* or *pending* (with the obvious meaning). After each step of inference, some of the proof values may change. The control procedure *propagates* the proof values in a bottom-up manner, changing *pending* into *proved* or *failed* when possible.

The mechanism that handles meta-variables splits on two levels — the level of a basic prover (in this presentation, the basic prover for predicate logic) and the level of the control procedure. Indeed, the basic prover *assigns* some particular values to meta-variables, while the control procedure insures that values assigned to a meta-variable on different branches of the deduction tree are *compatible*.

The behavior of the **basic prover** w.r.t. meta-variables is described in detail in Section 4. Using an example we illustrate the necessary behavior of the **control procedure** in order to handle correctly the proof search in the presence of meta-variables. Consider the deduction shown in Figure 1: After introducing a new meta-variable and splitting the goal, on the left branch we obtain a proof situation which can be transformed into a final one by the substitutions $\xi \mapsto a$ and $\xi \mapsto b$. At this point we do not know which substitution is correct, so we generate an OR node. A similar situation occurs on the right branch. A possible approach is to expand the deduction tree in the (expensive) breadth-first manner and try to combine the suggested substitutions. However, we proceed depth-first as shown in Figure 1. Nodes (ii) and (iii) have proof-values *pending* and will be expanded only after the control procedure detects that the currently computed substitutions do not solve the proof problem. Our approach is to collect the suggested substitutions bottom-up in the tree as soon as they appear, even if some alternative branches have not been examined yet. In fact, this technique can be combined with any tree-search strategy.

4 Basic Prover for First-order Logic

In the *Theorema* system, the basic prover that carries-out the inferences in first-order logic is called PND (Proof by Natural Deduction) [11]. In this section we describe the extension of this prover for handling metavariables. The reader may note that any set of inference rules can be used, either for predicate logic proving or for proving in some special domains. Additionally, the specificity of the domain can also be reflected in the particular unifier which is used (in the procedure described below and in the metavariables algorithm presented in Section 5), e.g. AC-unification for the domain of integers etc.

PND takes as parameters a proof situation $\Gamma \vdash G$ and the local context. Among other facts, PND keeps in the local context a set of restrictions of the form (29) and pairs of formulas.

At first, PND checks whether the given proof situation can be transformed to a final proof situation by substituting terms instead of metavariables. This is true in the following cases:

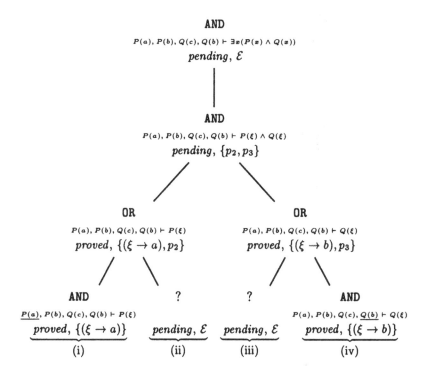

Figure 1. Unfinished deduction tree.

- There exists a unifier for the goal G and one of assumptions A_i. Let C_1 be G and C_2 be A_i.

- There exist two assumptions A_i and A_j, such that A_i and $\neg A_j$ have a unifier. Let C_1 be A_i and C_2 be $\neg A_j$.

If the pair (C_1, C_2) does not occur in the local proof context and a most general unifier σ for C_1 and C_2 satisfies the set of restrictions that is kept in the local proof context, the output of PND is the deduction tree consisting of an OR node with two successors. The first is the AND node with the proof value *proved*. The node is supplied with the unifier σ.

The second successor is the AND node with the proof value *pending*. The pair (C_1, C_2) is added to the local context, so this proof situation will never again be considered as a final because of (C_1, C_2).

Informally, the prover tries a substitution and keeps the possibility to go further if it is not correct. The proof situation does not change.

If the given proof situation cannot be transformed into a final proof situation, then PND tries to apply other rules, which basically split either the goal or one of the assumptions in a natural way. Some of these rules will create AND nodes with several branches (e. g. "prove conjunction by proving each subformula"). Of particular interest to the meta-variables method are the following:

- Rules that introduce Skolem constants:

$$\frac{\Gamma \vdash [G]_{x_1 \to c_1, \ldots, x_n \to c_n}}{\Gamma \vdash \forall x_1, \ldots, x_n G} ; \quad \frac{\Gamma, [F]_{x_1 \to c_1, \ldots, x_n \to c_n} \vdash G}{\Gamma, \exists x_1, \ldots, x_n F \vdash G},$$

  where $c_1, \ldots c_n$ are Skolem constants new to the deduction tree. For each meta-variable ξ that occurs in G (or F) the input proof situation we add to the local proof context the following restrictions

$$c_1 \notin \xi, \ldots, c_n \notin \xi.$$

  These restrictions guarantee that after replacing the meta-variables with their values, the Skolem constants do not occur in the quantified formulas.

- Rules that introduce meta-variables:

$$\frac{\Gamma \vdash [G]_{x_1 \to \xi_1, \ldots, x_n \to \xi_n} \vee \exists x_1, \ldots, x_n G}{\Gamma \vdash \exists x_1, \ldots, x_n G} ;$$

$$\frac{\Gamma, [A]_{x_1 \to \xi_1, \ldots, x_n \to \xi_n} \wedge \forall x_1, \ldots, x_n A \vdash G}{\Gamma, \forall x_1, \ldots, x_n A \vdash G},$$

  where $\xi_1, \ldots, \xi_n$ are new meta-variables. For each quantified variable y such that x_j is in the range of the quantifier over y, we add to the local proof context the following restriction

$$y \notin \xi_j.$$

  This restriction guarantees that ξ_j is free for x_j.

  Note that we do not need to update restrictions once they were created.

5   Algorithm

We consider first the situation of deterministic proofs — that is, we have only AND nodes. During the construction of the pre-proof, each node is associated with a set R of restrictions of the form (29), which are propagated top-down using the local context. Each final node (leaf) is assigned the proof value *proved* and the appropriate substitutions for the meta-variables. When all the subnodes of an AND node are *proved*, then we try to combine the corresponding substitutions and restrictions as shown below. If successful, the node is also *proved*. Inductively, this will construct bottom-up a correct substitution σ for the root.

The induction is provided by the function:

Function Simple_Combine

Input: A list of pairs $\{(\lambda_1, R_1), (\lambda_2, R_2), \ldots, (\lambda_n, R_n)\}$ [substitutions and restrictions of the subnodes].

Output: A pair (σ_N, R_N) or NULL [substitution and restrictions of the father node].

Method:

- Let $\lambda_i = \{\xi_{i,k} \to t_{i,k}\}_{k=1,\ldots}$.

- Find a most general unifier σ_N for $\langle \xi_{1,1} \ldots \xi_{n,1} \ldots \rangle$ and $\langle t_{1,1} \ldots t_{n,1} \ldots \rangle$.

- If σ_N exists and satisfies $\cup_i R_i$, then return $(\sigma_N, \cup_i R_i)$ else return NULL.

Obviously the resulting λ is the most general substitution mapping the corresponding subtree into a real proof.

Note that the unification algorithm used above can be particularly tailored to the special domains involved in the proof problem. Moreover, the unification can take into account the equalities that are in the assumption set of the proof situation (these equalities can be propagated together with the substitutions). Note also that the set of simultaneous substitutions may be seen as a system of equations which should be solved by a special constraint solver for the particular domain of the variables.

Consider now the case, when the deduction tree may contain OR nodes. An OR node may have some *pending* subnodes, as well as some *proved* subnodes $N_1, \ldots, N_n$ assigned with $(\lambda_1, R_1), \ldots, (\lambda_n, R_n)$. It is natural to assign to N the set $\{(\lambda_1, R_1), \ldots, (\lambda_n, R_n)\}$ of "possible" solutions.

The *pending* subnodes have no solutions assigned, however we keep a list of positions to them, in order to return there in case of failure.

Now we are ready to describe the general case. To each node we add a *system*—a set whose elements are either pairs of the form (σ_N, R_N) or pointers to nodes of the constructed deduction tree. Let functions `Pairs()` and `Pointers()` choose from a system its parts of the first and the second type respectively. A node is *proved* if the system assigned to it contains a pair. Thus, systems play the key role in manipulating proof values, in fact, we compute proof values and systems by means of one function `Compute_Value`.

To describe the function `Compute_Value` we need two auxiliary functions `CombineOR` and `CombineAND` that compute a system associated with `OR` and `AND` nodes of the deduction tree.

For an `OR` node the system is the union of systems of its immediate successors.

Function CombineOR

Input: A list of systems $\{S_1, \ldots, S_r\}$.

Output: A system S.

Method:

- Return the union of $S_1, \ldots, S_r$.

If all immediate successors of an `AND` node are proved, a system is assigned to each of them. Let denote these systems by $S_1, \ldots, S_r$. Each substitution occurring in `Pairs(`S_i`)` is correct for the corresponding subtree. `CombineAND` chooses an element s_i of `Pairs(`S_i`)` for all $i : 1 \leq i \leq r$, and passes $s_1, \ldots, s_r$ to the function `Simple_Combine`. The output of `Simple_Combine` contains a correct substitution or is NULL.

If no system can be obtained in this way, the function returns the union of sets of pointers occurring in the input systems. The proof search procedure can expand further the deduction tree starting from these nodes, then new leaves with the *proved* proof value may appear.

If there are no pointers, there is no hope to prove the node, the function returns NULL.

Function CombineAND

Input: A list of systems $\{S_1, \ldots, S_r\}$.

Output: A system $\{(\sigma_1, R_1), \ldots, (\sigma_M, R_M), p_1, \ldots, p_N\}$ or a list of pointers $\{p_1, \ldots, p_N\}$ or NULL.

Method:

- Let $S = \emptyset$.
- Let $P = \texttt{Pointers}(S_1) \cup \cdots \cup \texttt{Pointers}(S_r)$.
- For all sequences $(s_1, s_2, \ldots, s_r) \in \texttt{Pairs}(S_1) \times \texttt{Pairs}(S_2) \times \cdots \times \texttt{Pairs}(S_r)$
    - if $\texttt{Simple\_Combine}(s_1, \ldots, s_r) \neq \texttt{NULL}$, add the output of $\texttt{Simple\_Combine}(s_1, \ldots, s_r)$ to S.
- If $S = \emptyset$ and $P = \emptyset$, return $\texttt{NULL}$,
- else if $S = \emptyset$, return P,
- else return $S \cup P$.

The function $\texttt{Compute\_Value}$ combines the computation of proof values with the computation of systems. In addition, it builds the global list $\texttt{NList}$ of nodes.

Function $\texttt{Compute\_Value}$

Input: A node N having immediate successors $N_1, \ldots, N_t$.

Output: A proof value and a system assigned to N.

Side effect: Changes the global list $\texttt{NList}$.

Method:

- If N is an $\texttt{AND}$ node
    - If the proof value of one of $N_1, \ldots, N_t$ is *failed*, set the proof value of N to *failed* and exit.
    - If the proof value of one of $N_1, \ldots, N_t$ is *pending*, set the proof value of N to *pending* and exit.
    - Let $S_1, \ldots, S_t$ be the systems associated with $N_1, \ldots, N_t$ respectively. Let $S_{\texttt{AND}} = \texttt{CombineAND}(S_1, \ldots, S_t)$.
        * if $\texttt{Pairs}(S_{\texttt{AND}}) = \emptyset$ and $\texttt{Pointers}(S_{\texttt{AND}}) = \emptyset$, then set the proof value of N to *failed* and exit.
        * if $\texttt{Pairs}(S_{\texttt{AND}}) = \emptyset$ and $\texttt{Pointers}(S_{\texttt{AND}}) \neq \emptyset$, then set the proof value of N to *pending*, add $\texttt{Pointers}(S_{\texttt{AND}})$ to $\texttt{NList}$ and exit.
        * if $\texttt{Pairs}(S_{\texttt{AND}}) \neq \emptyset$, then set the proof value of N to *proved*, associate with N the system $S_{\texttt{AND}}$ and exit.

- If N is an OR node
  - If the proof values of all of $N_1, \ldots, N_t$ is *failed*, set the proof value of N to *failed* and exit.
  - If the proof values of all of $N_1, \ldots, N_t$ is *pending*, set the proof value of N to *pending* and exit.
  - If the proof value of one of $N_1, \ldots, N_t$ is *proved*,
    * Let $N'_{i_1}, \ldots, N'_{i_j}$ be those of the nodes $N_1, \ldots, N_t$ that have the *proved* proof value. Let $S_{i_1}, \ldots, S_{i_j}$ be the systems associated with $N'_{i_1}, \ldots, N'_{i_j}$ respectively.
    * Let $N''_{k_1}, \ldots, N''_{k_l}$ be those of the nodes $N_1, \ldots, N_t$ that have the *pending* proof value. Let $p_{k_1}, \ldots p_{k_l}$ be the pointers to the nodes $N''_{k_1}, \ldots, N''_{k_l}$ respectively. Let T be the system $\{p_{k_1}, \ldots p_{k_l}\}$.
    * Set the proof value of N to *proved*, associate with it the system $\texttt{CombineOR}(S_{i_1}, \ldots, S_{i_j}, T)$ and exit.

Some basic provers (as PND) assign systems to nodes. For compatibility, if a basic prover assigns to a node just a proof value, we associate with this node the empty system $\mathcal{E} = \{\epsilon, \emptyset\}$, where ϵ in the empty substitution. The empty system can be combined with any system.

The proof search is performed by the function $\texttt{Proof\_Search}$.

Function $\texttt{Proof\_Search}$

Input: Initial proof situation $\Gamma \vdash G$.

Output: If $\Gamma \vdash G$ is valid, a deduction tree T presenting its proof.

Method:

- Construct the deduction tree T consisting of a single AND node corresponding to the initial proof situation.
- Let $\texttt{NList} = \emptyset$.
- While the proof value of the root of T is *pending* do
  - Choose a leaf L of T having the *pending* proof value such that one of the following conditions is satisfied:
    1. no predecessor of L has the *proved* proof value;
    2. L is a successor of one of the nodes occurring in $\texttt{NList}$.

    The choice of L is determined by the search strategy.

- If no basic prover is applicable to L, set the proof value of L to *failed*.
- If only one basic prover $\mathcal{P}_i$ is applicable to L, replace L with the output of $\mathcal{P}_i$.
- If several basic provers $\mathcal{P}_{j_1}, \ldots, \mathcal{P}_{j_l}$ are applicable to L, form an OR node whose successors are generated by $\mathcal{P}_{j_1}, \ldots, \mathcal{P}_{j_l}$. Replace L with the newly formed tree.
- Propagate proof values from leaves to root applying inductively the function Compute_Value. During the propagation, the list NList may have changed.
- Exclude from the list NList all nodes that do not have successors with the pending proof value.

In Figure 1 (page 165) we presented a deduction tree where we show the proof values and the associated systems.

In the figure, p_2 and p_3 are pointers to the leaves (ii) and (iii) respectively. At the presented moment, the list NList contains nodes (ii) and (iii). Hence, the procedure chooses as L one of the leaves (ii) and (iii), w.l.g., the leaf (ii). After the procedure expands the deduction tree, we get the deduction tree presented in Figure 2.

The system associated with the root of the tree contains a correct substitution $\xi \to b$. The pointer p_3 shows the place where the deduction tree can be further expanded.

6 Improvements and Further Research

For the sake of simplicity, in this paper meta-variables may get values only in final nodes. Sometimes it is more natural to assign values also in intermediate nodes, e.g. for the Modus Ponens rule:

$$\frac{(\Gamma, A, B \vdash G)\sigma}{\Gamma, A, A' \implies B \vdash G},$$

where $A\sigma = A'\sigma$. Such rules can be easily incorporated into our framework and they are currently under implementation.

The developed implementation of the meta-variable method is efficient and does not depend on the search strategy. Note that using the same technique one can implement any method where the provability of a proof situation depends on extra data that has to be computed going bottom-up along a deduction tree.

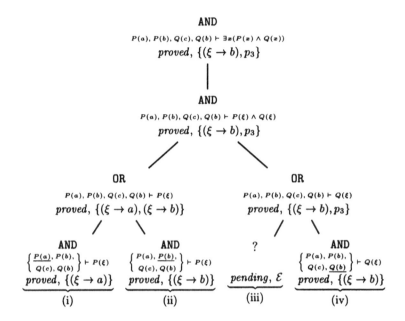

Figure 2. Finished deduction tree.

Our experiments show that an unrestricted introduction of meta-variables (as described in Section 4) leads to large deduction trees. Therefore it is needed to further investigate different search strategies and implement heuristics which allow to speed-up the proof search and reduce the search space. For instance, we are currently experimenting with techniques to guide the introduction of the meta-variables using the information from the proof situation. Additionally, we are introducing more special rules (e.g. instantiation of the goal if it matches some ground assumption), which are applied before the introduction of the meta-variables.

Acknowledgements

We would like to thank the referees for many helpful suggestions for improving the paper.

References

[1]   R. Hähnle and P. Schmitt. The liberalized δ-rule in free variable semantic tableaux. *Journal of Automated Reasoning*, 13(2):211–221, 1994.

[2] P. Baumgartner, N. Eisinger, and U. Furbach. A confluent connection calculus. In *CADE99*, pages 329–343, 1999.

[3] B. Beckert and R. Hähnle. Analytic tableaux. In W. Bibel and P. Schmitt, editors, *Automated Deduction—a Bases for Applications*, number 8 in Applied Logic Series, pages 11–41. Kluwer Academic Publishers, 1998.

[4] Bernhard Beckert. Depth-first proof search without backtracking for free variable semantic tableaux. In Neil Murray, editor, *Position Papers, International Conference on Automated Reasoning with Analytic Tableaux and Related Methods, Saratoga Springs, NY, USA*, Technical Report 99-1, pages 33–54. Institute for Programming and Logics, Department of Computer Science, University at Albany – SUNY, 1999.

[5] M. Beeson. Automatic derivation of the irrationality of *e*. In *CALCULEMUS 99*, Electronic Notes in Theoretical Computer Science, 99. www.elsevier.nl/locate/entcs.

[6] W. Bibel and P. Schmitt, editors. *Automated Deduction—a Bases for Applications*, Volume 1–3 of *Applied Logic Series*. Kluver Academic Publishers, 1998.

[7] K. Broda. The relationship between semantic tableau and resolution theorem proving. In *Proceedings of Workshop on Logic*, Debrecen, Hungary, 1980. Also published as technical report, Imperial College, Department of Computing, London, UK.

[8] F. M. Brown. Toward the automation of set theory and its logic. *Artificial Intelligence*, 10:281–316, 1978.

[9] B. Buchberger. Using *Mathematica* for doing simple mathematical proofs. In *4-th International Mathematica users' conference, Tokyo*. Wolfram Media Publishing, 1996.

[10] B. Buchberger, C. Dupre, T. Jebelean, F. Kriftner, K. Nakagawa, D. Vasaru, and W. Windsteiger. Theorema: A progress report. In *Calculemus*, 2000.

[11] B. Buchberger and T. Jebelean. The predicate logic prover. In *Proceedings of the First International Theorema Workshop*, number 97-20 in RISC Linz report series. Research Institute for Symbolic Computation, June 1997.

[12] B. Buchberger, T. Jebelean, F. Kriftner, M. Marin, E. Tomuta, and D. Vasaru. A survey of the *Theorema* project. In W. Kuechlin, editor, *Proceedings of ISSAC'97 (International Symposium on Symbolic and Algebraic Computation)*, Maui, Hawaii, July 1997.

[13] B. Buchberger and E. Tomuta. Combining provers in the *theorema* system. Technical report 98-02, Research Institute for Symbolic Computation, 1998.

[14] A.I. Degtyarev, A.V. Lyaletski, and M.K. Morokhovets. Evidence algorithm and sequent logical inference search. In *LPAR 1999*, pages 44–61, 1999.

[15] M. Fitting. *First-Order Logic and Automated Theorem Proving*. Springer, second edition, 1996.

[16] G. Gentzen. Untersuchungen über das logische schließ. *Mathematische Zeitschrift*, 39:176–210, 405–433, 1934.

[17] C. Kanger. *A simplified proof method for elementary logic*, pages 87–93. Computer Programming and Formal Systems / Studies in Logic and Foundations of Mathematics. North-Holland, 1963.

[18] J. Krajíček and P. Pudlák. The number of proof lines and size of proofs in first order logic. *Arch. Math. Logic*, 27:69–84, 1988.

[19] E. Melis and V. Sorge. Employing external reasoners in proof planning. In *CALCULEMUS 99*, Electronic Notes in Theoretical Computer Science, 99. www.elsevier.nl/locate/entcs.

[20] V. P. Orevkov. *Complexity of Proofs and Their Transformations in Axiomatic Theories*. Translations of Mathematical Monographs. American Mathematical Society, 1991.

[21] D. Prawitz. An improved proof procedure, 1960. Theoria 26.

[22] H. Wang. Toward mechanical mathematics. *IBM J. of Research and Develpment*, 4(1), 1960.

Exploring Properties of Residue Classes

Andreas Meier   Volker Sorge*

Abstract. *We report on an experiment in exploring properties of residue classes over the integers with the combined effort of a multi-strategy proof planner and two computer algebra systems. An exploration module classifies a given set and a given operation in terms of the algebraic structure they form. It then calls the proof planner to prove or refute simple properties of the operation. Moreover, we use different proof planning strategies to implement various proving techniques: from naive testing of all possible cases to elaborate techniques of equational reasoning and reduction to known cases.*

1 Introduction

We report on an experiment in exploring properties of operations on subsets of residue classes over the integers. The work was done within the scope of a project on an interactive mathematical course in algebra. For tutoring purposes it is necessary to have a large class of examples and counter-examples available to illustrate the difference of notions like group, monoid, etc. The examples also need to be suitable to teach different proving techniques. However, our work only provides the raw data that needs to be further processed to be usable within a tutor system. This process as well as how the examples are suitably presented to students are not topic of this paper.

We have chosen residue classes over the integers as a problem domain for the following reasons: Since operations on residue classes are relatively easy to handle automatically and intuitive to explain to a user they are ideal to exemplify some important concepts in algebra, such as associativity, unit elements, etc. Moreover, they provide a large number of possible examples and counter examples for the different concepts. A user of the interactive course can experiment with constructs such as $(\mathbb{Z}_3, x\bar{*}y), (\mathbb{Z}_3\backslash\{\bar{1}_3\}, (x\bar{+}x)\bar{*}y), \dots$. These are automatically classified in

---

*The author's work was supported by the 'Studienstiftung des deutschen Volkes'.

terms of algebraic structures, i.e., whether they form a magma, semi-group, quasi-group, monoid, loop, or group, by constructing the appropriate proofs which can in turn be used to guide the user in his experiments.

Technically, we have realized our goals by exploiting the infrastructure of the ΩMEGA system [6]. We use an exploration module that constructs proof obligations with respect to a given set and operation. These proof obligations are theorems of the form: the set is closed under the operation, or the operation is not associative, etc. Proof obligations are passed to ΩMEGA's multi-strategy proof planner MULTI [12] that constructs a proof with the help of the two computer algebra systems MAPLE [14] and GAP [4]. Although the proofs are not very pretentious in their own right they need to be constructed in a way that different proving techniques can be taught to a user of the interactive course. This is achieved by supplying distinct strategies to the proof planner so resulting plans reflect different proof approaches. In particular, we have implemented three different planning strategies: a very naive one testing all possible cases, a more elaborate one using as much as possible equational reasoning, and a third one whose goal is to apply already known theorems. We conducted a large number of experiments to test both the effectiveness and robustness of our proof planning approach.

The paper is organized as follows: Section 2 introduces the problem domain in more detail and Section 3 gives an account of multi-strategy proof planning in ΩMEGA. Section 4 explains how computer algebra aids during the proof planning process, Section 5 elaborates on the different proof strategies that realize different proving techniques, Section 6 reports on the exploration module employed to generate proof obligations, and Section 7 gives an account of the experiments we have conducted with the implemented machinery. We conclude by discussing some related work and hinting at some future work.

2 The Problem Domain

In this section we describe the kind of problems we are interested in together with the notation we will use throughout the remainder of the paper.

We are interested in classifying residue class sets over the integers together with given binary operations in terms of their basic algebraic properties. A residue class set over the integers is either the set of all congruence classes modulo an integer n, i.e., $\mathbb{Z}_n$, or an arbitrary subset of $\mathbb{Z}_n$. Concretely, we will be dealing with sets of the form $\mathbb{Z}_3, \mathbb{Z}_5, \mathbb{Z}_3 \backslash \{\bar{1}_3\}, \mathbb{Z}_5 \backslash \{\bar{0}_5\}$, $\{\bar{1}_6, \bar{3}_6, \bar{5}_6\}, \ldots$ where $\bar{1}_3$ denotes the congruence class 1 modulo 3. If c is an integer we write also $cl_n(c)$ for the congruence class c modulo n. A binary operation $\circ$ on a residue class set is given in λ-function notation. $\circ$ can be of the form $\lambda xy.x$, $\lambda xy.y$, $\lambda xy.c$ where c is a constant congruence class (e.g., $\bar{1}_3$), $\lambda xy.x\bar{+}y$, $\lambda xy.x\bar{*}y$, $\lambda xy.x\bar{-}y$, where $\bar{+}$, $\bar{*}$, $\bar{-}$ denote addition, multiplication, and subtraction on congruence classes over the integers respectively. Furthermore, $\circ$ can be any combination of the basic operations with respect to a common modulo factor, e.g., $\lambda xy.(x\bar{+}\bar{1}_3)\bar{-}(y\bar{+}\bar{2}_3)$.

Given a residue class set RS_n modulo n and a binary operation $\circ$ with respect to the same modulo factor n we try to establish or refute the following properties:

1. **Closure:** RS_n is closed under $\circ$. This is formalized by the defined concept $closed(RS_n, \circ)$ that abbreviates $\forall x{:}RS_n{.}\forall y{:}RS_n{.}(x \circ y) \in RS_n$. Here the variables x and y are of sort RS_n, i.e., the universal quantifiers range over the finite domain RS_n.

2. **Associativity:** RS_n is associative with respect to $\circ$. ($assoc(RS_n, \circ)$ $\equiv \forall x{:}RS_n{.}\forall y{:}RS_n{.}\forall z{:}RS_n{.}x \circ (y \circ z) = (x \circ y) \circ z$.)

3. **Unit element:** There exists a unit element in RS_n with respect to $\circ$. ($\exists e{:}RS_n{.}unit(RS_n, \circ, e) \equiv \exists e{:}RS_n{.}\forall y{:}RS_n{.}(y \circ e = y) \wedge (e \circ y = y)$.)

4. **Inverses:** Every element in RS_n has an inverse element with respect to $\circ$ and the unit element e. ($inverse(RS_n, \circ, e) \equiv \forall x{:}RS_n{.}\exists y{:}RS_n{.}(x \circ y = e) \wedge (y \circ x = e)$.)

5. **Divisors:** For every two elements of RS_n there exist two corresponding divisors in RS_n with respect to $\circ$. ($divisors(RS_n, \circ) \equiv \forall a{:}RS_n{.}\forall b{:}RS_n{.}(\exists x{:}RS_n{.}a \circ x = b) \wedge (\exists y{:}RS_n{.}y \circ a = b)$.)

Hence, we classify $(RS_n, \circ)$ with regard to the algebraic structure it forms, i.e., whether it is a magma (property 1 holds), semi-group (1+2), monoid (1+2+3), quasi-group (1+5), loop (1+5+3), or group (1+2+3+4)[1].

3 Multi-Strategy Proof Planning

Proof planning [2] considers mathematical theorems as planning problems where an *initial partial plan* is composed of the proof *assumptions* and the theorem as *open goal*. A proof is then constructed with the help of abstract planning steps, called *methods*, that are essentially partial specifications of tactics known from tactical theorem proving.

In the ΩMEGA system [6] the traditional proof planning approach is enriched by incorporating mathematical knowledge into the planning process

---

[1]Naturally property 5 holds for a group as well.

(see [13] for details). That is, methods can encode general proving steps as
well as knowledge particular to a mathematical domain. Moreover, *control
rules* specify how to traverse the search space by preferring, rejecting, or
enforcing the application of methods in certain domains or proof situations.
ΩMEGA's new proof planner, MULTI [12], allows also for the specification
of different planning strategies to control the overall planning behavior.
Proof plans in ΩMEGA result in proof objects in a variant of the natural
deduction calculus [5]. We will present them in this paper in a linearized
style as introduced in [1]. A proof line is of the form $L.\ \Delta\vdash F\ (\mathcal{R})$, where
L is a unique label, $\Delta\vdash F$ a sequent denoting that the formula F can be
derived from the set of hypotheses Δ, and $(\mathcal{R})$ is a justification expressing
how the line was derived.

Methods are generally represented in a declarative way as for instance
∃IResclass given below, a method specific for our problem domain. Its
purpose is to instantiate an existentially quantified variable over a residue
class set with a witness term for which a certain property P holds and to
reduce the initial statement on residue classes to a statement on integers.
The witness term has to be a concrete element of the residue class set. However, if the method is applied in an early stage of the proof, the planner
generally has no knowledge on the true nature of the witness term. Therefore, the method invokes a middle-out-reasoning [10] process to postpone
the actual instantiation, that is, a meta-variable is used as placeholder for
the actual witness term which will be determined at a later point in the
planning process and subsequently substituted.

Method: ∃IResclass			
Premises	$\oplus L_3, \oplus L_1$		
Appl. Cond.	$ResclassSet(RS_n, n, N_{set})$		
Conclusions	$\ominus L_5$		
Declarative Content	(L_1) Δ	$\vdash mv \in N_{set}$	(open)
	(L_2) Δ	$\vdash c \in RS_n$	(ConResclSet L_1)
	(L_3) Δ	$\vdash P[cl_n(mv)]$	(open)
	(L_4) Δ	$\vdash P[c]$	(ConRescl L_3)
	(L_5) Δ	$\vdash \exists x{:}_{RS_n}\!.P[x]$	(∃sortI $L_2\ L_4$)

∃IResclass is given in terms of the original goal, the *conclusion* L_5, the
two new open goals it produces, the *premises* L_1 and L_3, and the inference steps deriving L_5 from L_1 and L_3, given in the *declarative content*.
The method is applicable during the planning process if the current planning goal can be matched against the formula of L_5 and if additionally
the *application conditions (Appl. Cond.)* are satisfied. The condition

$ResclassSet(RS_n, n, N_{set})$ is fulfilled if RS_n, the sort of the quantified variable x, qualifies as a residue class set of the form given in Section 2. Its successful evaluation binds the method variables n and N_{set} to the modulo factor of RS_n and the set of integers corresponding to the congruence classes of RS_n, respectively. For instance, the evaluation of $ResclassSet(\mathbb{Z}_2, n, N_{set})$ yields $n \mapsto 2$ and $N_{set} \mapsto \{0, 1\}$. The necessary inference steps are indicated by the justifications $ConResclSet$ and $ConRescl$ in lines L_2 and L_4. A congruence class with respect to the modulo factor n is denoted as $cl_n(mv)$ in line L_3 where mv is the introduced meta-variable.

To influence the planners behavior control-rules can be specified and are evaluated at certain choice points, e.g., the selection of a planning goal, or of an applicable method, or an instantiation of a meta-variable. An example of a control-rule used in our context is `TryAndErrorStandardSelect`:

```
(control-rule TryAndErrorStandardSelect
     (kind methods)
     (IF (disjunction-supports S))
     (THEN (select (∀IResclass
                    ConCongCl
                    (∨E** () (S))
                    ∃IResclass))))
```

The control-rule is evaluated before a method is chosen by the planner. It states that if the current goal is supported by a disjunctive support line S the application of the methods `∀IResclass`, `ConCongCl`, `∨E**`, and `∃IResclass` is attempted in this order. `∀IResclass` and `ConCongCl` are domain-specific methods where the latter converts statements on residue classes into corresponding statements on integers. The former reduces goals containing a universal quantification over a residue class set similar to `∃IResclass`. On the contrary, `∨E**` is not a domain-specific method. When applied it performs a case split with respect to a set of disjunctive supports of a goal.

Different problem solving behaviors are realized using the multi-strategy extension MULTI [12] of the conventional proof planner. Essentially *strategies* are used to parameterize certain algorithms of the planner. A planning algorithm like forward- or backward-planning can be supplied with termination criteria and a certain set of methods and control rules to determine its behavior. One strategy we developed in the context of residue classes is a parameterization of the standard planning algorithm *PPlanner*:

Strategy: TryAndError		
Appl-Cond	ResidueClassGroupProperty	
Algorithm	*PPlanner*	
Parameters	Methods	∀IResclass, ConCongCl, ∨E**, ∃IResclass, ...
	C-Rules	TryAndErrorStandardSelect, ...
	Termin.	No-Subgoal

According to its application conditions (Appl-Cond) **TryAndError** can be applied to goals stating one of the properties given in Section 2. It employs the standard planning algorithm *PPlanner* with a certain set of methods and a certain set of control-rules. The strategy either fails or terminates when its initial goal is fully justified. The application and the selection of the different strategies can also be influenced by *strategic control-rules* which — similar to control-rules about methods — prefer, reject, or enforce the application of strategies in certain situations.

4 Using Computer Algebra

In this section we describe how we employ symbolic calculations to guide and simplify the search for proof plans. In particular, we use the mainstream computer algebra system MAPLE [14] and GAP [4], a system specialized on algebra. We will not be concerned with the technical side and the soundness issues of integrating computer algebra with proof planning in general and with ΩMEGA in particular since this has already been discussed in [16, 9]. We rather concentrate on the cooperation between the systems in the context of exploring residue class properties.

We use symbolic calculations in two ways: (1) in control rules hints are computed to help guiding the planning process, and (2) within method-applications equations are solved with MAPLE to simplify the proof. As side-effect both cases can restrict possible instantiations of meta-variables.

(1) is implemented in the control rule select-instance which is used in the TryAndError strategy. The rule is triggered after decomposition of an existentially quantified goal which results in the introduction of a meta-variable as substitute for the actual witness term. After an existential quantifier is eliminated the control rule computes a hint with respect to the remaining goal that is used as a restriction for the introduced meta-variable. For instance, when proving that the residue class set RS_n is not closed under the operation ∘, i.e., that there exist $a, b \in RS_n$ such that $a \circ b \notin RS_n$, the control rule would supply a hint as to what a and b might be. If hints can be computed the meta-variables are instantiated before the proof planning proceeds.

The hint system is implemented as follows: We have implemented a small routine in ΩMEGA that constructs a multiplication table with respect to the given set and operation. It is employed to check the closure property and to check the existence of divisors, i.e., the axiom for quasi-groups. Furthermore, if the computed multiplication table is closed under the respective operation it is used to construct the appropriate magma in GAP. GAP can then be employed to query for associativity and to compute the unit element and inverses for the single elements. Most query functions return useful results in both the positive and the negative case: That is, for instance, if GAP can compute a unit element for a given magma this element is returned. In case GAP fails to find a unit element the multiplication table is used to determine a set of elements that suffice to refute the existence of a unit element for the given magma. A special case is the failure of the query for associativity, since then we try to use MAPLE to compute a particular solution for the associativity equation. If such a non-general solution exists it is exploited to determine a triple of elements for which associativity does not hold.

Use (2) of calculations is realized within the `Solve-Equation` method. Its purpose is to justify an equational goal using MAPLE and to instantiate meta-variables if necessary. In detail, it works as follows: if an open goal is an equation MAPLE's function `solve` is applied to check whether the equality actually holds. If the equation contains meta-variables these are considered as the variables the equation is to be solved for and they are supplied to `solve` as additional arguments. In case the equation involves modulo functions with the same factor on both sides MAPLE's function `msolve` is used, instead.

5 Using Different Strategies

To be able to demonstrate different proving techniques we employ three different strategies for the MULTI proof planner. In this section we elaborate those strategies using examples to point out the major differences while trying to avoid the tedious details.

5.1 Exhaustive Case Analysis

The motivation for the first strategy, called `TryAndError`, is to implement a rather naive approach of proving a property of a residue class set. It proceeds by rewriting statements on residue classes into corresponding statements on integers, especially by transforming the residue class set into a set of corresponding integers. It then exhaustively checks all possible

L_1.	L_1	$\vdash c_1' \in \mathbb{Z}_2$	(Hyp)
L_2.	L_1	$\vdash c \in \{0,1\}$	$(ConResclSet\ L_1)$
L_3.	L_3	$\vdash c = 0$	(Hyp)

$$\vdots$$

L_{12}.	L_1, L_3	$\vdash \exists y{:}\mathbb{Z}_2{\bullet}(cl_2(c)\bar{+}y{=}\bar{0}_2) \wedge (y\bar{+}cl_2(c){=}\bar{0}_2)$	$(\exists IResclass\ L_{11}\ L_{10})$
L_{13}.	L_{13}	$\vdash c = 1$	(Hyp)
L_{14}.	L_1, L_{13}	$\vdash 0 = 0$	$(Solve-Equation)$
L_{15}.	L_1, L_{13}	$\vdash mv \in \{0,1\}$	$(\vee IR\ L_{14})$
L_{16}.	L_1, L_{13}	$\vdash 0 = 0$	$(Solve-Equation)$
L_{17}.	L_1, L_{13}	$\vdash 0 = 0$	$(Solve-Equation)$
L_{18}.	L_1, L_{13}	$\vdash (1 + c)\ mod\ 2 = 0\ mod\ 2$	$(SimplNum\ L_{13}\ L_{16})$
L_{19}.	L_1, L_{13}	$\vdash (c + 1)\ mod\ 2 = 0\ mod\ 2$	$(SimplNum\ L_{13}\ L_{17})$
L_{20}.	L_1, L_{13}	$\vdash (c + 1)\ mod\ 2 = 0\ mod\ 2\ \wedge$	$(\wedge I\ L_{18}\ L_{19})$
		$\qquad (1 + c)\ mod\ 2 = 0\ mod\ 2$	
L_{21}.	L_1, L_{13}	$\vdash (cl_2(c)\bar{+}cl_2(mv)) = \bar{0}_2) \wedge$	$(ConCongCl\ L_{20})$
		$\qquad (cl_2(mv)\bar{+}cl_2(c) = \bar{0}_2)$	
L_{22}.	L_1, L_{13}	$\vdash \exists y{:}\mathbb{Z}_2{\bullet}(cl_2(c)\bar{+}y{=}\bar{0}_2) \wedge (y\bar{+}cl_2(c){=}\bar{0}_2)$	$(\exists IResclass\ L_{21}\ L_{15})$
L_{23}.	L_1	$\vdash \exists y{:}\mathbb{Z}_2{\bullet}(cl_2(c)\bar{+}y{=}\bar{0}_2) \wedge (y\bar{+}cl_2(c){=}\bar{0}_2)$	$(\vee E{**}\ L_2\ L_{12}\ L_{22})$
L_{24}.		$\vdash \forall x{:}\mathbb{Z}_2{\bullet}\exists y{:}\mathbb{Z}_2{\bullet}(x\bar{+}y = \bar{0}_2) \wedge (y\bar{+}x = \bar{0}_2)$	$(\forall IResclass\ L_{23})$
L_{25}.		$\vdash inverse(\mathbb{Z}_2, \lambda xy.x\bar{+}y, \bar{0}_2)$	$(DefnCon\ L_{24})$

combinations of these integers with respect to the property that is to be proved or refuted. TryAndError proceeds in two different ways, depending on whether (1) a universal or (2) an existential property has to be proved. Both cases can be observed in the example proof of the statement that $\mathbb{Z}_2$ has inverses with respect to the operation $\lambda xy.x\bar{+}y$ and the unit element $\bar{0}_2$, given above.

In case (1) a split over all the elements in the set involved is performed and the property is proved for every single element separately. We observe this in the proof of the universally quantified formula in line L_{24}. An application of the method $\forall$IResclass to L_{24} yields the lines L_{23}, L_1, and L_2. The disjunction contained in L_2 ($c \in \{0,1\}$ can be viewed as $c = 0 \vee c = 1$) triggers the first case split with the application of $\vee$E**. Subsequently MULTI tries to prove the goal in line L_{23} twice (in the lines L_{12} and L_{22}), once assuming $c = 0$ (in line L_3) and once assuming $c = 1$ (in line L_{13}).

In case (2) the single elements of the set involved are examined until one is found for which the property in question holds. In our example proof this is, for instance, done after the application of the method $\exists$IResclass to L_{22} yielding the lines L_{15} and L_{21}. The case analysis is then performed by successively choosing different possible values for mv with the $\vee$IR, $\vee$IL methods — in our example mv is either 0 or 1 as given in line L_{15}. For a selected instantiation MULTI can then either finish the proof or it backtracks to test the next instantiation. To minimize this search the `TryAndError` strategy enables MULTI to invoke the `select-instance` control-rule after the application of ∃IResclass. As described in Section 4 `select-instance` can compute a hint on the likely instantiation for the meta-variable mv it is used as instantiation for the very first case. If no hint can be computed or the proof fails despite the hint MULTI proceeds with its conventional search.

After eliminating all quantifiers and performing all possible case splits the `TryAndError` strategy reduces all remaining statements on residue and congruence classes to statements on integers. These are then solved by numerical simplification and equational reasoning.

5.2 Equational Reasoning

The aim of the second strategy, called `EquSolve`, is to use as much as possible equational reasoning to prove properties of residue classes. Similarly to the `TryAndError` strategy it converts statements on residue classes into corresponding statements on integers. But instead of then checking the validity of the statements for all possible cases, it tries to solve occurring equations in a general way. The strategy can be successfully applied to prove associativity, the existence of a unit element, inverses and divisors, only. We observe its approach with a proof of the same example theorem $inverse(\mathbb{Z}_2, \lambda xy.x \bar{+} y, \bar{0}_2)$ as in Subsection 5.1:

$L_1. \quad L_1 \vdash c_1' \in \mathbb{Z}_2$ $\qquad\qquad\qquad\qquad$ (Hyp)

$L_2. \quad L_1 \vdash c \in \{0, 1\}$ $\qquad\qquad\qquad\qquad$ $(ConResclSet\ L_1)$

$L_{15}. L_1 \vdash mv \in \{0, 1\}$ $\qquad\qquad\qquad\qquad$ $(Open)$

$L_{18}. L_1 \vdash (mv + c)\ mod\ 2 = 0\ mod\ 2$ $\qquad$ $(Solve\!-\!Equation\{mv \leftarrow c\})$

$L_{19}. L_1 \vdash (c + mv)\ mod\ 2 = 0\ mod\ 2$ $\qquad$ $(Solve\!-\!Equation)$

$L_{20}. L_1 \vdash (c + mv)\ mod\ 2 = 0\ mod\ 2\ \wedge$ $\qquad$ $(\wedge I\ L_{19}\ L_{18})$
$\qquad\qquad\quad (mv + c)\ mod\ 2 = 0\ mod\ 2$

$L_{21}. L_1 \vdash (cl_2(c)\bar{+}cl_2(mv)=\bar{0}_2)\ \wedge$ $\qquad\qquad$ $(ConCongCl\ L_{20})$
$\qquad\qquad\quad (cl_2(mv)\bar{+}cl_2(c)=\bar{0}_2)$

$L_{22}. L_1 \vdash \exists y{:}\mathbb{Z}_2{\scriptstyle\blacksquare}((cl_2(c)\bar{+}y = \bar{0}_2) \wedge (y\bar{+}cl_2(c) = \bar{0}_2))$ $\quad$ $(\exists IResclass\ L_{21}L_{15})$

$L_{24}. \quad \vdash \forall x{:}\mathbb{Z}_2{\scriptstyle\blacksquare}\exists y{:}\mathbb{Z}_2{\scriptstyle\blacksquare}((x\bar{+}y = \bar{0}_2) \wedge (y\bar{+}x = \bar{0}_2))$ $\quad$ $(\forall IResclass\ L_{23})$

$L_{25}. \quad \vdash inverse(\mathbb{Z}_2, \lambda xy.x\bar{+}y, \bar{0}_2)$ $\qquad\qquad\qquad$ $(DefnCon\ L_{24})$

The construction of the proof is in the beginning (lines L_{25} through L_{20}) nearly analogous to the one in the preceding section. The only exception is that no case splits are carried out after the applications of ∀IResclass and ∃IResclass. Instead we get two equations in the lines L_{18} and L_{19}

which can be generally solved using the `Solve-Equation` method. As described in Section 4 this method uses MAPLE to compute a general solution and possibly substitutes meta-variables. In the case of our example the meta-variable mv is substituted by c during the first application of `Solve-Equation`. This is indicated by $\{mv \leftarrow c\}$ in the justification of line L_{18}. However, this substitution is not carried out directly during the proof planning process but merely added as a constraint on the meta-variable. Nevertheless, from there on the proof planner treats any occurrence of mv as if it were substituted. Once a complete proof plan is constructed all computed substitutions for meta-variables are applied. This treatment of meta-variable substitutions simplifies the backtracking procedure of the proof planner.

The constraint on mv changes the formula implicitly in the remaining open goal L_{15} to $c \in \{0, 1\}$. Since the naive testing of the possible cases fails (neither $c = 0$ nor $c = 1$ hold immediately), the goal is reduced to $closed(\mathbb{Z}_2, \lambda xy.x)$, an assertion about a residue class property. This problem is then returned to MULTI which in turn selects and applies again strategies to this goal.

5.3 Applying Known Theorems

The motivation for our third strategy `ReduceToSpecial` is to incorporate the application of already proved theorems. Contrary to `TryAndError` and `EquSolve` whose idea is to reduce expressions on residue classes to expressions on numbers, the `ReduceToSpecial` strategy tries to tackle new problems by applying already known theorems. Theorems are stored in ΩMEGA's theory database and, in order to keep the number of possible theorems small, only those in the theory of residue classes are eligible for application.

`ReduceToSpecial` uses two methods to apply theorems: the primary method is `Apply-Assertion` that applies theorems that match an open goal. To ensure termination `Apply-Assertion` uses first-order matching with α-equality on λ-abstractions, only. `Apply-Assertion` is a general method to apply arbitrary theorems and is not tailored for particular problem domains and it is, in particular, not always sufficient for our examples. Therefore, it is necessary to have other, domain-specific methods that are able to apply theorems in a more complex way. So far, we have implemented one such method, `ReduceClosed`, that is specialized to apply theorems containing statements on closure properties of residue classes.

We observe the behavior of the `ReduceToSpecial` strategy with the following proof for the theorem $closed(\mathbb{Z}_5, \lambda x, y.(x \bar{*} y) \bar{+} \bar{3}_5)$:

L_3.	$\vdash \bar{3}_5 \in \mathbb{Z}_5$	(*InResclSet*)
L_4.	$\vdash 5 \in \mathbb{Z}$	(*InInt*)
L_5.	$\vdash closed(\mathbb{Z}_5, \lambda xy_\bullet x)$	(*ApplyAss ClosedFV*)
L_6.	$\vdash closed(\mathbb{Z}_5, \lambda xy_\bullet y)$	(*ApplyAss ClosedSV*)
L_7.	$\vdash 5 \in \mathbb{Z}$	(*InInt*)
L_8.	$\vdash closed(\mathbb{Z}_5, \lambda xy_\bullet \bar{3}_5)$	(*ApplyAss ClosedConst L$_3$*)
L_9.	$\vdash closed(\mathbb{Z}_5, \lambda xy_\bullet x \bar{*} y)$	(*ReduceClosed ClComp $\bar{*}$ L$_4$ L$_5$ L$_6$*)
L_{10}.	$\vdash closed(\mathbb{Z}_5, \lambda xy_\bullet (x \bar{*} y) \bar{+} \bar{3}_5)$	(*ReduceClosed ClComp $\bar{+}$ L$_7$ L$_8$ L$_9$*)

The following are the theorems involved (stated informally)[2]:

1. Each residue class set RS_n is closed with respect to the operations: $\lambda xy_\bullet c$ if $c \in RS_n$ (corresponding to the theorem *ClosedConst*), $\lambda xy_\bullet x$ (*ClosedFV*), and $\lambda xy_\bullet y$ (*ClosedSV*).

2. Each complete residue class set $\mathbb{Z}_n$ which is closed under the binary operations op_1 and op_2, is also closed under the composed binary operation $\lambda xy_\bullet (x \; op_1 \; y) \circ (x \; op_2 \; y)$ where $\circ \in \{+, -, \bar{*}\}$ (*ClComp$\bar{+}$, ClComp$\bar{-}$, ClComp$\bar{*}$*).

While the theorems under (1) are applied by `Apply-Assertion`, the theorems under (2) need to be applied with `ReduceClosed`. This is due to the fact that when, for instance, applying the theorem

$$\forall n : \mathbb{Z}_\bullet \forall op_1 \bullet \forall op_2 \bullet (closed(\mathbb{Z}_n, op_1) \wedge closed(\mathbb{Z}_n, op_2)) \Rightarrow$$
$$closed(\mathbb{Z}_n, \lambda x, y_\bullet (x \; op_1 \; y) \bar{+} (x \; op_2 \; y))$$

to line L_{10} the necessary instantiations for the operations have to be $op_1 = \lambda xy_\bullet x \bar{*} y$ and $op_2 = \lambda xy_\bullet \bar{3}_5$. If the theorems under (2) were treated in a general way we would need higher order matching to compute the instantiations and thus `Apply-Assertion` is not sufficient. Instead, we implemented a special decidable matching algorithm employed by the `ReduceClosed` method. The algorithm is suitable to handle the cases under (2) and when the method is applied the premises of the respective theorem involved will result in new open sub-goals.

Some other methods associated with the strategy `ReduceToSpecial` are `InInt` which closes goals of the form $n \in \mathbb{Z}$ if n is an integer and `InResclSet` closing goals of the form $c \in RS_n$ if c is an element of the residue class set RS_n.

We have also experimented with bookkeeping already solved problems and trying to reduce new problems to these. However, this approach is not feasible since for large sets of problems the comparison of a new problem with those already solved is rather expensive.

---

[2]Similarly, our database contains theorems suitable for associativity, unit element, inverses, and divisors problems.

5.4   Comparison of the Three Strategies

Only the `TryAndError` strategy is applicable to all possible occurring prob-
lems. And, judging from the experiments, it also always successfully derives
a proof, even when the hints provided by the computer algebra systems fail.
The strategy `EquSolve` is just suitable for proving the properties associativity, the existence of a unit element, of inverse elements, and of divisors.
It can neither be used to refute any of these properties nor to prove closure.
Moreover, it sometimes even fails for problems it can be applied to when
MAPLE returns facts useless in our context (e.g., a term involving a rational number). The `ReduceToSpecial` strategy can currently be applied to
closure, associativity, unit element, inverses, and divisors problems. Up to
now we have no theorems for negated problems in our database. The strategy is, however, quite limited since it can only succeed in those special cases
that are covered by the given theorems. Therefore, MULTI has a strategic
control rule that tries `ReduceToSpecial`, `EquSolve`, and `TryAndError` in
this order.

From a performance point of view, `TryAndError` is generally the most
time consuming of the three strategies and delivers the longest proof plans
since it painstakingly wades through all possible cases. Both disadvan-
tages increase the larger the involved residue class sets are. Contrary, both
`EquSolve` and `ReduceToSpecial` are relatively fast and yield rather lean
plans. Moreover, they are independent of the size of the involved sets. However, with a growing number of possible theorems to apply, the efficiency
of `ReduceToSpecial` strategy might also decrease.

As far as generality of the involved strategies is concerned, one should
note that both `TryAndError` and `EquSolve` possess only three domain specific methods: ∀IResclass, ConCongCl, and ∃IResclass. All their other
methods are domain independent and used also for proof planning in other
domains. Except for `Apply-Assertion` all the methods of the strategy
`ReduceToSpecial` are constructed for the strategy. Hence, only a few of
these will be useful in other domains.

6   Exploring Properties

We are currently working with a module that enables us to automatically
explore the single properties of residue classes. The motivation for this is
that a user only has to supply a residue class set and an operation and the
module will stepwise construct theorems to either prove or refute single
properties. The final result stating the nature of the given algebra is then
derived automatically. The module also aids the automatic exploration of
large testbeds as described in Section 7.

The input to the exploration module is a set and a corresponding operation. To classify the given algebra it employs the same functionality as described in Section 4, i.e., it uses the query function for the single properties. It step-by-step queries about properties, generates theorems according to the obtained results and passes them to MULTI to prove. Unfortunately, the results of the query functions cannot be used in the proof planning process since MULTI can only use information given in the form of methods, strategies, and control rules. However, this is not of great weight since the repeated computations are relatively inexpensive.

In detail the exploration works as follows: It first checks with the help of the multiplication table whether the operation is closed on the given set. If not, the appropriate theorem is generated, passed to MULTI and if proved successfully the exploration stops. If the set is indeed closed under the operation and the according theorem is proved by MULTI we have at least a magma and carry on with the exploration. The next exploration steps work similar, that is, the exploration module generates theorems to prove that a property either holds or that it does not hold.

After establishing that we have at least a magma, the next property checked is associativity to see if we have a semi-group. If this holds, the module checks if there exists a unit element, i.e., whether we have a monoid. Finally, if inverses exist we have a group. In case the check for associativity fails, the exploration module proceeds by checking the existence of divisors to see if we have a quasi-group. And if the given set together with the operation really forms a quasi-group the next check is again for the existence of a unit element, meaning we would have a loop. The exploration will then stop in any case since we already know that the given algebra does not form a group (since it is not associative).

7 Experiments

To design and implement the approach we describe in this paper and especially to extract the necessary methods and control rules we used a total of 21 examples. To show the appropriateness of the methods and control rules and to check the effectivity of the implemented strategies they need to be tested against a large number of examples that differ from those used during the design process. Moreover, considering the purpose of the implementation, i.e., its use in a tutor system where a user can pose virtually any residue class (at least up to a certain order) together with an arbitrary combination of operations as problem, it is essential to test the robustness of the implementation with a wide range of examples.

We are currently working with a testbed of roughly 13 million auto-matically generated examples. These examples are constructed from the possible subsets of the residue classes modulo n where n is in between 2 and 10 together with operations that are systematically constructed from the basic operations. We tried to exclude repeating and trivial cases, e.g., sets with only one element or operations like $x - x$, as far as possible. From the testbed we have explored 4152 examples, mainly involving $\mathbb{Z}_2$ to $\mathbb{Z}_6$ and some of their subsets. As interesting cases we have encountered 661 magmas, 990 semi-groups, 186 quasi-groups, 22 monoids, and 261 groups so far. However, these figures do not indicate whether the classified alge-braic entities are actually distinct since we do not check for isomorphism between the elements in the testbed yet. This is subject of future work. When proving the constructed proof obligations MULTI could successfully employ ReduceToSpecial to a sample of 15% and EquSolve to a different set accounting for another 15% of the examples. The remaining 70% of the examples could only be solved by the TryAndError strategy. How-ever, these figures are not as disappointing as they seem at a first glance considering that nearly all proofs involving the closure property of non-complete residue class sets (i.e., sets such as $\mathbb{Z}_3 \backslash \{\bar{1}_3\}$) and the refutation of properties could only be solved with the TryAndError strategy.

Additionally, we are experimenting with a set of 100 non-theorems. Although most of these can be successfully detected as false, this generally takes a long time. We are currently testing ways to minimize this time. Furthermore, the methods we developed for proving the group properties of the residue classes proved general enough to tackle also commutativity of single operations and distributivity properties of two operations (only the hint system for the strategy TryAndError had to be slightly expanded). We tested this with 10 examples so far.

8 Related Work and Conclusion

We have reported on an experiment in exploring properties of residue classes over integers. Single properties are step-wise checked by an ex-ploration module that generates the appropriate proof obligations. These proof obligations are passed to a multi-strategy proof planner to be proved. The proof planner can draw on different planning strategies in order to solve the problems. We employ the computer algebra systems GAP and MAPLE to guide and simplify both the exploration and the proof planning process. We gave some empirical evidence that the currently implemented methods form a robust set of planning operators that suffices to explore the domain of residue classes.

There are various accounts on experiments of combining computer algebra and theorem proving in the literature (see [8] for just a few). However, they generally deal with the technical and architectural aspects of those integrations as well as with correctness issues and not with application of the combined system to a specific problem domain. A possibly fruitful cooperation between the deduction system Nuprl and the computer algebra system Weyl in the domain of abstract algebra is sketched in [7]. More concrete work in exploration in finite algebra is reported in [3, 11, 15] where model generation techniques are used to tackle quasi-group existence problems. In particular, some open problems in quasi-group theory were solved.

Our work has neither the claim to advance the theoretical and architectural aspects of the integration of deduction and computer algebra systems nor do we expect to make any new discoveries. It rather concentrates on having a robust and flexible machinery for working with examples in the context of tutor systems. Moreover, it shows how a combination of various systems can be fruitfully employed to large classes of examples. Although one major ingredient in our setup is ΩMEGA's multi-strategy proof planner, the presented work does not reflect the possible power of proof planning in general, since the problem solutions are still quite algorithmic and the necessary search is rather limited.

Future work will include to extend the exploration module so we can classify residue classes with two associated operations in terms of rings, integral domains and fields. Moreover, we are currently also implementing appropriate methods and strategies to reason about isomorphism between the residue classes both to detect isomorphic structures among those already classified and to have examples for the interactive course in algebra for the concept of isomorphism.

References

[1] P. Andrews. Transforming matings into natural deduction proofs. In *Proc. of CADE-5*, p. 281–292, 1980.

[2] A. Bundy. The use of explicit plans to guide inductive proofs. In *Proc. of CADE-9*, volume 310 of *LNCS*, p. 111–120. Springer, 1988.

[3] M. Fujita, J. Slaney, and F. Bennett. Automatic generation of some results in finite algebra. In *Proc. of IJCAI'93*, p. 52–57. Morgan Kaufmann, 1993.

[4] The GAP Group, Aachen, St Andrews. *GAP – Groups, Algorithms, and Programming, Version 4*, 1998. http://www-gap.dcs.st-and.ac.uk/~gap.

[5] G. Gentzen. Untersuchungen über das Logische Schließen I und II. *Mathematische Zeitschrift*, 39:176–210, 405–431, 1935.

[6] The ΩMEGA Group. ΩMega: Towards a Mathematical Assistant. In *Proc. of CADE-14*, volume 1249 of *LNAI*, p. 252–255. Springer, 1997.

[7] P. Jackson. Exploring Abstract Algebra in Constructive Type Theory. In *Proc. of CADE-12*, volume 814 of *LNCS*, p. 590–604. Springer, 1994.

[8] D. Kapur and D. Wang, editors. *Journal of Automated Reasoning— Special Issue on the Integration of Deduction and Symbolic Computation Systems*, volume 21(3). Kluwer Academic Publisher, 1998.

[9] M. Kerber, M. Kohlhase, and V. Sorge. Integrating Computer Algebra Into Proof Planning. In [8], p. 327–355.

[10] I. Kraan, D. Basin, and A. Bundy. Middle-out reasoning for logic program synthesis. Tech. Report MPI-I-93-214, MPI, Saarbrücken, Germany, 1993.

[11] W. McCune. A Davis-Putnam program and its application to finite first-order model search: Quasigroup existence problems. Technical Memorandum ANL/MCS-TM-194, Argonne National Laboratory, USA, 1994.

[12] E. Melis and A. Meier. Proof planning with multiple strategies. In *Proc. of the First International Conference on Computational Logic*, 2000. Springer.

[13] E. Melis and J. Siekmann. Knowledge-based proof planning. *Artificial Intelligence*, 1999.

[14] Darren Redfern. *The Maple Handbook: Maple V Release 5*. Springer, 1999.

[15] J. Slaney, M. Fujita, and M. Stickel. Automated reasoning and exhaustive search: Quasigroup existence problems. *Computers and Mathematics with Applications*, 29:115–132, 1995.

[16] Volker Sorge. Non-Trivial Computations in Proof Planning. In *Proc. of FroCoS 2000*, volume 1794 of *LNCS*. Springer, 2000.

Defining Power Series and Polynomials in Mizar

Piotr Rudnicki*   Christoph Schwarzweller

Andrzej Trybulec†

Abstract. *We report on the construction of formal multivariate power series and polynomials in the Mizar system. First, we present how the algebraic structures are handled and how we inherited the past developments from the Mizar library. The Mizar library evolves and past contributions are revised and (usually) generalized. Our work on formal power series caused a number of such revisions. It seems that revising past developments with an intent to generalize them is a necessity when building a data base of formalized mathematics. And this poses a question: how much generalization is best?*

1   Introduction

Mathematics, especially algebra, uses dozens of structures; groups, rings, vector spaces, to name few of the most basic ones. These structures are closely connected to each other giving raise to inheritance. For example, each ring is a group with respect to its addition and hence every theorem about groups trivially holds for rings also. Furthermore there is a trend towards introducing more general structures: semi-rings as a generalization of rings, modules as a generalization of vector spaces, etc. Again, theorems about a structure are trivially true for any structure derived from it. The derived structure inherits everything from its ancestors.

In mechanized proof-checking systems the issues of inheritance have to be made explicit. It is by far non trivial to build a proof-checker for which

---

*Supported by NSERC Grant OGP9207.

†Supported by NSERC Grant OGP9207 and NATO CRG 951368.

theorems for groups apply also to rings. Generalizations, as mentioned above, may result in building a sizeable graph of inheritance and only extensive practice can say how good is a particular solution. The issue is further complicated by inertia induced through the existing developments in a proof-checking environment. On the one hand, one would like to inherit as much as possible from the past, on the other hand one wants to change the past if it turns out to be inconvenient for the task at hand. And the task at hand is usually too big to start everything from scratch.

In this paper we describe the construction of formal multivariate power series and polynomials in the Mizar system [4], during which we had to deal with these problems. We discuss the tools Mizar offers to build algebraic structures; tools, we believe, providing a flexible mechanism allowing the kind of inheritance omnipresent in mathematics. Generalization is a more complex task. Of course one can easily derive rings from semi-rings; however, there is a challenge when the rings have been already introduced in the past and one aims at introducing semi-rings. The question is what to do with the theorems about rings already proven and stored in a library. A number of them will hold for semi-rings also. Stating and proving them again would not only be a tedious job but would blow up the library. Alternatively, the library has to be revised as a whole. We discuss some issues of generalizations and library revisions.

2   Defining Algebraic Domains in Mizar

The Mizar construction of formal multivariate polynomials aimed at defining the ring of polynomials over a minimal algebraic domain which would permit such a construction. The definition of an algebraic domain is founded on a structure mode providing the primitive notions, and next, the axioms for a specific class of structures are defined as properties of the underlying structure mode.

In our case, the structure mode of interest is `doubleLoopStr`, defined in [3]. Figure 1 illustrates the relationship of `doubleLoopStr` to other structure modes[1]. The bottom definition introduces the following constructors:

- The structure mode `doubleLoopStr`, that may be used to qualify variables, e.g. `let S be doubleLoopStr` or form predicates, e.g. `T is doubleLoopStr`. However, T for which the above holds, may have other fields besides those listed in the definition, if the type of T is derived from `doubleLoopStr`.

---

[1]See [1], [8] and [3] to learn more about these structures.

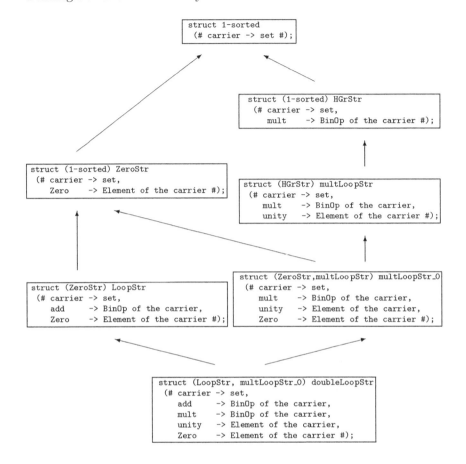

Figure 1. Derivation of `doubleLoopStr`.

- The attribute `strict` which when used as `strict doubleLoopStr` gives the type of structures that have no additional fields besides the ones mentioned in the definition. The attribute symbol `strict` is heavily overloaded as every definition of a structure mode defines a new attribute denoted by this symbol.

- The aggregate functor that is used to construct aggregates of the form `doubleLoopStr(#c,a,m,u,z#)`, where c is a set, a and m binary operations on c, u and z two fixed elements of c. Structures denoted by aggregates are `strict`.

- The forgetful functor which when used as the `doubleLoopStr` of S creates a strict structure from S (provided S has a type widening to `doubleLoopStr`). This functor denotes the aggregate:

```
doubleLoopStr (# the carrier of S,
               the add of S, the mult of S,
               the unity of S, the Zero of S #)
```

If S is `strict doubleLoopStr`, then S = the `doubleLoopStr` of S.

The mode `doubleLoopStr` is derived from `LoopStr` and `multLoopStr_0`. This means that type `doubleLoopStr` widens to both, or in other words is a subtype of both `LoopStr` and `multLoopStr_0`.

Typically, a structure definition introduces also some selector functors to access its fields. The selector functors are introduced in the first structure definition in which the selector is used. The structure mode `1-sorted` defines the selector functor `the carrier of`. It may be used for any `1-sorted` structure, e.g. `ZeroStr`, `LoopStr`, `doubleLoopStr`. The selector functor `the Zero of` is introduced by `ZeroStr` and the selector functor `the add of` by `LoopStr`. In the case of `multLoopStr_0` no new selectors are introduced, `mult` and `unity` are inherited from `HGrStr` and `multLoopStr`, respectively. In the case of `doubleLoopStr` also all selector functors are inherited.

`ZeroStr` is a common ancestor of `LoopStr` and `multLoopStr_0`. In this way we ensure that `carrier` and `Zero` are the same in both. The definition of `ZeroStr` introduces `Zero` as a new selector, `carrier` is inherited from `1-sorted` that is common ancestor for most structures in the Mizar library (MML).

If S is defined to satisfy S = doubleLoopStr(#c,a,m,u,z#) then

```
the 1-sorted of S = 1-sorted(#c#)
the ZeroStr of S = ZeroStr(#c,z#)
the LoopStr of S = LoopStr(#c,a,z#)
the multLoopStr_0 of S = multLoopStr_0(#c,m,u,z#)
the doubleLoopStr of S = doubleLoopStr(#c,a,m,u,z#)
```

that is the `doubleLoopStr` of S = S.

The order of selectors in a structure definition serves syntactic purposes only and it can be chosen arbitrarily (with obvious restriction that a selector s_1 that occurs in the type of a selector s_2 must be put before s_2). The structures in Mizar are not tuples but rather partial functions on selectors and selectors must not be identified with just a place in the aggregate functor.

A structure mode defines a backbone on which algebraic domains are built. The desired properties of an algebraic domain are introduced usually one at a time through defining appropriate attributes. For example, associativity of addition is formulated ([8]) as:

```
definition
let S be non empty LoopStr;
attr S is add-associative means
  for a,b,c being Element of the carrier of S
  holds (a + b) + c = a + (b + c);
end;
```

(a + b is a shorter notation for (the add of S).[a,b]; this notation is usually defined right after the selector functor is introduced, see [8].) The attribute add-associative is defined for the structure mode LoopStr, in which the selector add is introduced. Through inheritance—the mode doubleLoopStr widens to the mode LoopStr as the latter is an ancestor of the former—the attribute is available for objects of mode doubleLoopStr.

Using separate attributes, various properties of algebraic domains are defined. These attributes can then be combined into clusters:

```
definition
cluster add-associative ri ght_zeroed ri ght_complementable
        Abelian commutative associative left_unital
        right_unital distributive Field-like non de generated
        (non empty doubleLoopStr);
existence
  proof
    Demonstrate the existence of an object with all listed attributes.
  end;
end;
```

The attributes in the cluster above were introduced for various structure modes, all inherited by doubleLoopStr. For example, empty is defined for 1-sorted, Abelian for LoopStr, commutative for HGrStr. The attribute distributive is defined for doubleLoopStr as it could not have been defined earlier.

The existence proof in the cluster definition is necessary to avoid empty modes that are not allowed in Mizar. Once we have proven the existence of an object with a cluster of attributes, we can introduce a mode of the desired algebraic domain:

```
definition
mode Field is
```

```
          add-associative right_zeroed right_complementable
          Abelian commutative associative left_unital
          right_unital distributive Field-like non de generated
          (non empty doubleLoopStr);
    end;
```

The mode `Field` is an abbreviation for a `doubleLoopStr` having the attributes given in its definition. Note that through inheritance the definition of `Field` just combines various notions; most of them exist on their own merit. A definition of a `Ring` could share the same backbone structure with `Field` and its attributes could be a subset of `Field`'s attributes. Therefore each theorem about a `Ring` would be applicable to a `Field`.

Conditional clusters is another Mizar mechanism through which we obtain reuse of theorems. A conditional cluster expresses the fact that a Mizar object that enjoys some attributes also enjoys another ones. For example, the rather trivial fact that a commutative binary operator with a right zero also possesses a left zero, can be expressed as follows:

```
    definition
    cluster Abelian right_zeroed -> left_zeroed (non em pty LoopStr);
    coherence
      proof
        Demonstrate that the desired implication holds.
      end;
    end;
```

Once the above conditional cluster has been registered, predicates and functions defined for `left_zeroed LoopStr` are now also available for all other objects whose type widens to `Abelian right_zeroed LoopStr`. Also, theorems proven for `left_zeroed LoopStr` are now applicable to `Abelian right_zeroed LoopStr` and all other types widening to it. Mizar checker tacitly processes all available conditional clusters and they are not explicitly referenced.

3   Formal Multivariate Power Series and Polynomials

The construction of formal power series and polynomials is presented in [5]. We build power series as functions from power products into a structure L of coefficients. A power product itself is a function, called `bag`, from a given set of variables into the natural numbers.

Variables are elements of an arbitrary set X. When we need the variables to be ordered, we use ordinals as X but we prefer to be as general as possible. A bag over a set of variables X is defined in terms of the concept of ManySortedSet [7].

```
definition
let X be set;
mode bag of X is
    natural-yielding finite-support ManySortedSet of X;
end;
```

The attribute natural-yielding means that the values of a bag are natural numbers, whereas finite-support describes the property of a function as having only finitely many values not equal to zero. The set of all bags of X is then defined and named Bags X.

Several operations on bags are defined, for example addition (b1 + b2) used for multiplying power products and restricted subtraction (b1 -' b2) used for dividing power products. Also, we introduced the usual order on power products and the concept of their divisibility.

Given a structure S, a formal power series over S with the variables in X assigns to each power product over X a coefficient, an element of S. Consequently a Series of S,X is a function from Bags X into S:

```
definition
let X be set, S be 1-sorted;
mode Series of X,S -> Function of (Ba gs X),S means
    not contradiction;
end;
```

Note that nothing is required from the structure S, in particular no addition over S has to be available. These assumptions are introduced later when necessary to ensure additional properties of series. For example, defining addition of series requires addition of the elements of S, hence is defined for LoopStr:

```
definition
let n be set, L be right_zeroed (non empty LoopStr),
    p,q be Series of n,L;
func p + q -> Series of n,L means
    for x being bag of n holds it.x = p.x + q.x;
end;
```

(Note: the attribute right_zeroed is not really needed and will be probably eliminated when the library is revised.)

Definition of multiplication required a bit more work: p*q on a bag b is obtained by considering all decompositions of b into bags b1 and b2 such that b = b1 + b2. This is done with the helper functor `decomp` which gives the finite sequence of decompositions of b ordered in increasing order of the first component. For this we required that the variables are identified with a certain ordinal:

```
definition
let n be Ordinal,
    L be add-associative right_complementable
        right_zeroed (non empty doubleLoopStr),
    p,q be Series of n, L;
func p*q -> Series of n, L means
  for b being bag of n
  ex s being FinSequence of the carrier of L
  st it.b = Σ s &
      len s = len decomp b &
      for k being Nat st k ∈ dom s
      ex b1, b2 being bag of n st π(decomp b,k) = <*b1, b2*> &
                                  π(s,k) = p.b1 · q.b2;
end;
```

(We would like to mention that proving the associativity of this multiplication presented a technical challenge.)

We also defined the operators p - q and -p with the obvious meaning as well as the zero series and the unit series denoted by 0_(n,L) and 1_(n,L).

Polynomials are a special case of formal power series; they are the series having only finitely many power products with non-zero coefficients, that is series with a finite support (written `finite-Support` to distinguish from `finite-support`, introduced earlier). Due to this restriction the underlying structure L must have a zero, hence be a `ZeroStr`.

```
definition
let n be Ordinal, L be non empty ZeroStr;
mode Polynomial of n,L is finite-Support Series of n,L
end;
```

All the functors defined for series and resulting in series can be applied to polynomials, however, the types of these functors are series and not polynomials. We have to explicitly state when performing operations on polynomials we obtain polynomials, that is that the resulting series has finite support. This problem is solved by employing functorial clusters, in which exactly this is stated (and proved), for example:

```
definition
let n be Ordinal, L be right_zeroed (non empty LoopStr),
```

```
      p,q be Polynomial of n,L;
 cluster p + q -> finite-Support;
 coherence
   proof
     Show that the result of adding two Polynomials has finite-Support.
   end
 end;
```

Putting it all together we get the ring of polynomials over a structure L as a doubleLoopStr in which the single components are identified with the corresponding just defined operators. Note that the underlying structure L is not a full commutative ring. We only used attributes necessary to ensure that the operators for polynomials again result in a polynomial.

```
 definition
 let n be Ordinal,
     L be right_zeroed add-associative right_complementable unital
         distributive non trivial (non empty doubleLoopStr);
 func Polynom-Ring(n,L) -> strict non empty doubleLoopStr means
   (for x being set
    holds x ∈ the carrier of it iff x is Polynomial of n,L) &
   (for x,y being Element of it, p,q being Polynomial of n,L
    st x = p & y = q holds x + y = p + q) &
   (for x,y being Element of it, p,q being Polynomial of n,L
    st x = p & y = q holds x · y = p * q) &
   0.it = 0_(n,L) &
   1_ it = 1_(n,L);
 end;
```

So far we only defined an instance of a doubleLoopStr; nothing is said about the usual algebraic properties of a polynomial ring. To constitute Polynom-Ring(n,L) as a ring, the necessary attributes are introduced in cluster registrations. For some of the attributes, additional properties of L were necessary. For example, it turned out that in order to prove the commutativity of multiplication, we also needed the addition of polynomials to be commutative.

```
 definition
 let n be Ordinal,
     L be Abelian add-associative right_zeroed
         right_complementable commutative unital distributive
         non trivial (non empty LoopStr);
 cluster Polynom-Ring(n,L) -> commutative;
 end;
```

Finally, to prove distributivity of `Polynom-Ring(n,L)` we had to use attributes implying that L is a ring with a unit, but not a commutative one.

```
definition
let n be Ordinal,
    L be right_zeroed Abelian add-associative
        right_complementable unital distributive associative
        non trivial (non empty doubleLoopStr);
cluster Polynom-Ring (n, L) -> unital right-distributive;
end;
```

4   Evaluating Multivariate Polynomials

The next natural step is to define the evaluation of polynomials in the underlying structure L [6]. To define the evaluation as a function from the ring of polynomials over L into L, it is not necessary for L to be a ring. But in order to prove that the evaluation of polynomials is a homomorphism, further properties of L are necessary, namely that L is a non trivial commutative ring with 1.

First, we resolve the problem of evaluating a power product b which is a bag of n.

```
definition
let n be Ordinal,
    b be bag of n,
    L be unital non trivial (non empty doubleLoopStr),
    x be Function of n,L;
func eval(b,x) -> Element of the carrier of L means
  ex y being FinSequence of the carrier of L
  st len y = len SgmX(RelIncl n, support b) &
     it = Π y &
     for i being Nat st 1 <= i & i <= len y holds
         y|.i = power(L).((x·SgmX(RelIncl n, support b))|.i,
                         (b·SgmX(RelIncl n, support b))|.i);
end;
```

We use a helper function x evaluating the variables n into L. To get the interesting part of x, that is an evaluation of the variables occurring with non-zero exponents in b, the functor SgmX is employed. This functor takes a finite set—here the support of a bag—and a linear order for the set and returns a finite sequence in which the elements of the set occur in increasing order. This sequence is then composed with x to get the finite sequence of values and with b to get the corresponding finite sequence of exponents.

The exponentiation is then performed pointwise yielding a finite sequence y of elements of L. The result of the evaluation of b with respect to x is the product of the values of y. We get this product using the functor Π which takes a finite sequence (over a structure allowing for elements occurring in this sequence [9]).

The structure L has to meet two requirements: the existence of a unity and that it is not trivial (has at least two elements). However, in order to prove that the evaluation respects multiplication of power products, that is to prove the rather obvious fact

```
eval(b1+b2,x) = eval(b1,x)  ·  eval(b2,x)
```

it turned out that L has to provide a commutative multiplication with a (left and right) unity. As this property is necessary to prove multiplicativity of evaluation, it follows that the evaluation of polynomials is a homomorphism only if the underlying structure L is a commutative ring with 1.

The definition of the evaluation of a polynomial is defined in analogous way to the evaluation of power products:

```
definition
let n be Ordinal,
    L be right_zeroed add-associative ri ght_complementable unital
            distributive non trivial (non em pty doubleLoopStr),
    p be Polynomial of n,L,
    x be Function of n,L;
func eval(p,x) -> Element of the carrier of L means
  ex y being FinSequence of the carrier of L
  st len y = len SgmX(BagOrder n, Support p) &
     it = Σ y &
     for i being Nat st 1 <= i & i <= len y holds
         y|.i = (p·SgmX(BagOrder n, Support p))|.i ·
                eval(((SgmX(BagOrder n, Support p))|.i),x);
end;
```

The next goal was to prove that the functor `eval` is a homomorphism from the polynomial ring over L into L. To do so we introduced a functor `Polynom-Ring(n,L,x)` taking an ordinal number, a structure L and variable evaluation function x as parameters, assigning to each polynomial p the value of `eval(p,x)`.

```
definition
let n be Ordinal,
    L be right_zeroed add-associative ri ght_complementable unital
            distributive non trivial (non em pty doubleLoopStr),
```

```
    x be Function of n, L;
 func Polynom-Evaluation(n,L,x) -> ma p of Polynom-Ring(n,L),L
    means for p being Polynomial of n,L holds it. p = eval(p,x);
 end;
```

Proving the properties of a homomorphism required additional assumptions concerning L, in particular, to prove that the evaluation of polynomials is compatible with the multiplication of polynomials we had to assume for the first time, that the underlying structure L is indeed a commutative ring with 1. Thus we ended up with the following

```
 definition
 let n be Ordinal,
     L be right_zeroed add-associative ri ght_complementable
         Abelian well-unital distributive non trivial
         commutative associative (non em pty doubleLoopStr),
     x be Function of n,L;
 cluster Pol ynom-Evaluation(n,L,x) -> Rin gHomomorphism;
 end;
```

5   Library Revisions

When starting to define polynomials, we wanted to keep the number of new definitions as small as possible. Hence we examined the Mizar Mathematical Library (MML) to find concepts we could use for our task. Several problems occurred.

We found out (not for the first time) that many basic theorems were missing. For example, the functor Σ sums up elements of a finite sequence; it is clear that if all but one particular element equal zero, the sum in fact is this element. This theorem had not been proven before.

Some concepts were introduced for a too specific structure, hence not general enough to use them in our case. For example, the functor **power**, for exponentiation with natural numbers, was defined for groups, whereas we wanted to use it in a structure providing only unity. Of course one can define the functor again for the more general case, but this does not seem appropriate in a library. The solution is to revise the MML, that means generalizing the original definition and reformulating the theorems concerning this concept.[2]

On the one hand this problem seems natural. If one is writing an article about for example groups in which one needs a functor—and one does not

---

[2]Sometimes it turns out that the proof of a theorem actually does not use all properties of the structure it is about.

find it in the MML—one simply defines it. And why should one think about more general solutions, if it works well for the theorems intended to prove? In addition, it is rather hard, if even possible, to estimate how general a definition should be in order to provide optimal benefit for future users of MML.

On the other hand, while proving theorems about the new concept, one usually observes which properties of the underlying structure are necessary to prove it and which are not. The correctness proof of the `power` functor, for example, definitely did not use the properties of a group. The same holds for defining formal power series and polynomials: we first did (and finished) this job for polynomials with a finite number of variables only, before we realized that we already had developed all the machinery for constructing power series in arbitrary number of variables.

Another point connected with this problem is that sometimes it may be better to be not as general as possible. For example, although one can build the theory of polynomials in one variable out of our approach, by using `Polynomial-Ring(1,R)`, this seems not to be the best solution. Doing so would require R to be a commutative and associative ring, just because these properties were necessary to prove multiplicativity of the evaluation in the general case. But it seems that for polynomials with one variable only this property can be established with weaker assumptions on R.[3] So the question remains: How much generalization is best?

6   Conclusions and Further Work

In this paper we described the construction of formal multivariate power series and polynomials in the Mizar system. The main concern was to present the possibilities Mizar offers to build algebraic structures. Although these possibilities are quite elegant and include inheritance in the usual mathematical style, library revisions were necessary during our work.

We plan to go on with our work on polynomials, among other goals is getting more insight into the way of dealing with inheritance in algebraic structures. We want to check how applications of our general theory of polynomials can be done. One of the goals is to develop a theory of polynomials over finite fields—as used in coding theory—thus to restrict the underlying structure of a polynomial ring. Also the theory of one-variable polynomials will be constructed separately, although it could be build using our approach. This may serve as a case study concerning the question how much generalization is best.

---

[3] At the moment the theory of polynomials in one variable is developed separately in order to check this.

Another plan is to work towards the theory of Groebner bases and Buchberger-like algorithms. For that it is necessary to develop the theory of ideals first. This seems to be an algebraic topic with many applications and hence could also contribute to solve the problems we discussed here.

References

[1] Association of Mizar Users, Library Committee, *Preliminaries to Structures*. Available on WWW: http://mizar.org/JFM/Addenda/struct_0.abs.html.

[2] Grzegorz Bancerek and Krzysztof Hryniewiecki, *Segments of Natural Numbers and Finite Sequences*. Formalized Mathematics, 1(1):107–114, 1990. Available on WWW: http://mizar.org/JFM/Vol1/finseq_1.abs.html.

[3] Eugeniusz Kusak, Wojciech Leończuk and Michał Muzalewski, *Abelian Groups, Fields and Vector Spaces*. Formalized Mathematics, 1(2):335–342, 1990. Available on WWW: http://mizar.org/JFM/Vol1/vectsp_1.abs.html.

[4] Piotr Rudnicki and Andrzej Trybulec. On Equivalents of Well-foundedness. An experiment in Mizar. *Journal of Automated Reasoning*, 23:197–234, 1999.

[5] Piotr Rudnicki and Andrzej Trybulec, *Multivariate Polynomials with arbitrary Number of Variables*. Formalized Mathematics, 8(1):317–332, 1999. Available on WWW: http://mizar.org/JFM/Vol11/polynom1.abs.html.

[6] Christoph Schwarzweller and Andrzej Trybulec, *Evaluation of Multivariate Polynomials*. To appear in *Formalized Mathematics*, 2000. Available on WWW: http://mizar.org/JFM/Vol12/polynom2.abs.html.

[7] Andrzej Trybulec, *Many-sorted sets*. Formalized Mathematics, 4(1):15–22, 1993. Available on WWW: \tthttp://mizar.org/JFM/Vol5/pboole.abs.html.

[8] Wojciech A. Trybulec, *Vectors in Real Linear Space*. Formalized Mathematics, 1(2):291–296, 1990. Available on WWW: http://mizar.org/JFM/Vol1/rlvect_1.abs.html.

[9] Wojciech A. Trybulec, *Lattice of Subgroups of a Group. Frattini Subgroup*. Formalized Mathematics 2(1):41–47, 1991. Available on WWW: http://mizar.org/JFM/Vol2/group_1.abs.html.

Logic and Dependent Types in the Aldor Computer Algebra System

Simon Thompson

Abstract. *We show how the Aldor type system can represent propositions of first-order logic, by means of the 'propositions as types' correspondence. The representation relies on type casts (using* `pretend`*) but can be viewed as a prototype implementation of a modified type system with* type evaluation *reported elsewhere [9]. The logic is used to provide an axiomatisation of a number of familiar Aldor categories as well as a type of vectors.*

1 Introduction

An earlier paper [9] outlined a new combination of symbolic computation and formal reasoning by means of an embedding of a logic in the type system of the computer algebra system Axiom [6] or, more specifically, its library compiler Aldor [12]. Such an embedding has two important consequences.

It can make the system *sound*, so that, for instance, $x^{n+1}/(n+1)$ is only used as the integral of x^n when n is not equal to -1. It can also make a CAS *more expressive* in combining logical steps with calculational ones. We illustrate this in Section 5 by giving axiomatisations of various algebraic structures within the CAS.

The literature contains a number of different strategies proposed for combining computer algebra and theorem proving; see, for instance, [2, 3, 1]. Our approach is distinctive in requiring minimal changes to an existing system; indeed, using the strategy outlined in this paper it can be done to an *unchanged* system.

The embedding is based on the *propositions as types* correspondence, whereby a logical proposition in a constructive logic is represented by a type in a programming language [7, 10]. Proofs of the proposition are then elements of the type in question. A more detailed exposition follows in the body of the paper, but as an illustration observe that a conjunction of two formulas is represented by the product of two types, each type representing one of the conjuncts; elements of this product will be pairs of elements which correspond to proofs of the two conjuncts.

To represent quantified formulas, it is necessary to use *dependent types*. A dependent function f is one for which the *type* of the result f(a) depends on the *value* of the argument a. A dependent function type represents a universal formula and a dependent record represents an existential statement. The dependent types of Aldor are thus crucial to the embedding.

To realise the proposed embedding, it is necessary to modify the way that the system handles dependent types; the earlier paper explains this in detail, but suffice it to say here that we require that *types are evaluated* in the same way that values are. To take an example which will be developed in the present paper, in the current Aldor system the vector types Vector(2+3) and Vector(5) are *different*, since the vector lengths 2+3 and 5 are left unevaluated; after evaluation they are, of course, the same.

This paper discusses how a *prototype* of the modifications can be implemented in Aldor using type casts using pretend. An expression (e pretend T) is accepted by the typechecker as having type T; typechecking and code-generation then proceed on this basis.

Our implementation is useful for a variety of reasons. First, it provides a *proof of concept* for our ideas. Secondly, the implementation *guides the design* of a modified type checker by illustrating the particular places that type evaluation and normalisation need to take place. Finally, it provides a *testbed* in which to explore the different ways that a logical view of mathematical objects can be integrated into an already existing CAS.

This work also shows that the code generated by the Aldor system for our technically type-incorrect programs is correct, and thus shows that the back end of the Aldor system can support code generation for our modified notion of type dependency without itself being changed.

The structure of the paper is as follows. Section 2 contains a brief introduction to some of the salient features of Aldor; others are introduced in the remainder of the paper. Sections 3 and 4 describe how propositional and predicate logic are represented in Aldor. Building on these, Section 5 shows how this framework can be used to axiomatise various algebraic structures in Aldor. Section 6 shows how a dependent type of vectors can be developed in Aldor and we make some concluding remarks in Section 7.

I am grateful to John Shackell, Leonid Timochouk and James Beaumont as well as Erik Poll and Thérèse Hardin and her group for discussions about this work. Stephen Watt patiently explained to me some of the finer details of Aldor. Martin Dunstan and two anonymous referees gave very helpful feedback on an earlier version of the paper. NAG Ltd. have kindly given us access to the Aldor compiler for research in this area.

2 An Introduction to Aldor

The core of Aldor [12] (also known in the past as AXIOM-XL and A$^\sharp$ [11]) is a strongly-typed functional language which has much in common with modern functional languages like SML and Haskell [5].

Since Aldor is designed with mathematics in mind, its type system is more complex than those of most programming languages. It supports *overloading* of symbols and *coercion* between types as well as permitting functions to have *dependent* types. Moreover the language allows an entity like the collection of integers to be seen in various different ways, depending on its context. For example, the integers might be seen as a set of values, a group, an integral domain, a subset of the real numbers and so forth. To do this, the language allows types and functions to be collected into abstract data types which are known as *domains* in Aldor.

The type of a domain, which is described by a signature, is called a *category*. Categories bear a strong resemblance to Java interfaces, and thus the effect of allowing a domain to belong to different categories is to support different views of the same structure. An example of a domain and its category is given in Section 3, which also contains a brief overview of the Aldor mechanisms supporting these definitions. Categories can be built on top of other categories, giving a version of inheritance between domains. Categories can also be parameterised by values including domains, since types are themselves values (of type `Type`). Because of this Aldor possesses a rich structure of interdependent program units.

Current descriptions of Aldor, [12, 11], give informal definitions of the type system; [8] gives a formal description of the essence of the Aldor type system. Examples of Aldor programs are given in remainder of this paper as well as in [9, 12]; features of the language are introduced as they are needed. The paper uses Aldor version 1.1.12p5 with the `axllib` library.

3 Propositional Logic

This section gives an overview of how propositional logic is implemented in Aldor. The logical connectives become constructors of types, and the logical rules to introduce and eliminate connectives similarly become functions over logical formulas.

Implication

The simplest connective to represent is implication, =>, which is represented
by the function type constructor, ->.

An implication is introduced by a function definition, and eliminated
by a function application. For example, a proof that a formula A implies
itself is given by the function

```
(x:A):A +-> x
```

The notation `(a:A):C +->` e denotes a function: in this case it is the
function which takes a value a of type A to the result e of type C.

Given proofs a of A and f of A->B, a proof of B is given by f(a). Other
examples will emerge in the course of discussion of the other connectives.

Conjunction

As was said in the introduction, conjunction can be represented by a product type; in Aldor this means that a `Record` is used. A direct representation
would be given by defining

```
And(A:Type,B:Type):Type == Record(fst:A,snd:B);
```

A function to introduce a conjunction would have type

```
andIntro(a:A,b:B):And(A,B)
```

since this takes proofs of the two conjuncts to build a proof of the conjunction. From a proof of a conjunction we may recover proofs of the two
conjuncts, so that there should be two function to *eliminate* a conjunction:

```
andElim1(p:And(A,B)):A
andElim2(p:And(A,B)):B
```

Instead of this direct approach, we define an Aldor *domain* to represent
conjunction; this approach is more in keeping with the idioms of Aldor; the
full definition appears in Figure 1.

The signature of the domain is given in the `with{ ... }`, where the
symbol % stands for 'the type being defined', that is `And(A,B)`. This part
of the definition gives the category of `And(A,B)`; the part following the keyword `add` gives the implementation of the domain itself. The representation
of the domain is given by `Rep`; note that it is necessary explicitly to import
from the representation, `Record(fst:A,snd:B)`, to allow the overloaded
record operations to be used at this particular type.

```
And(A:Type,B:Type)  : with{
        andIntro : (a:A,b:B)  -> %;
        andElim1 : (p:%)      -> A;
        andElim2 : (p:%)      -> B; }
    == add {
    Rep       == Record(fst:A,snd:B);
    import from Record(fst:A,snd:B);

    andIntro(a:A,b:B):% == per [a,b];
    andElim1(p:%):A      == (rep p).fst;
    andElim2(p:%):B      == (rep p).snd; }
```

Figure 1. The Aldor domain for conjunction: `And`.

Following this are the definitions of the functions themselves. In `andIntro` a record is built, and in the elimination functions record fields are selected. The functions `rep` and `per` are used to convert from the type being defined (`%`) to the representation chosen (`Rep`) and *vice versa*.

Given these functions one can begin to write proofs using the introduction and elimination rules. A proof of `B&A` from `A&B` is given by the function

```
flip(A:Type,B:Type,p:And(A,B)):And(B,A)
    == andIntro(andElim2(p),andElim1(p));
```

which takes the two components of the proof `p` of `And(A,B)` and 'flips' their order to give a proof of the 'flipped' formula, `And(B,A)`.

Disjunction

A disjunction is represented by a `Union` of two types, and so in defining the domain `Or(A:Type,B:Type)` the representation is given by

```
Rep == Union(inl:A,inr:B);
```

To introduce a proof of a disjunction it is sufficient to give a proof of either disjunct, and so there are two introduction rules, given by functions of type

```
orIntro1(a:A):%
orIntro2(b:B):%
```

(recall that `%` stands for `Or(A,B)` in this case.)

How is a disjunction eliminated? To prove an arbitrary formula `C` from `A\/B` it is enough to have proofs of `C` from `A` and `B` separately:

```
orElim(C:Type,f:A->C,g:B->C,p:%):C
   == {
   val == (rep p);                                    (1)
   if (val case inl)                                  (2)
      then f(val.inl)                                 (3)
      else g(val.inr)};
```

Converting the proof p of type `Or(A,B)` to its representation `val` (line (1)), a case switch is performed on `val` (line (2)). If `val` comes from the left half of the union (line (3)) then a proof of `C` is produced by applying `f` – a proof of `A->C` – to the proof of `A` contained in `val`; otherwise, `g` is used.

An example combining implication, conjunction and disjunction is a proof of the formula `(A=>C)&(B=>C)` from `((A\/B)=>C)`:

```
andOr(A:Type,B:Type,C:Type,p:Or(A,B)->C):And(A->C,B->C)
   ==
{ import from Or(A,B);   -- Needed to give the appropriate
                         -- meaning to orIntro1 and orIntro2.
   andIntro((a:A):C +-> p(orIntro1(a)),
           (b:B):C +-> p(orIntro2(b)));};
```

Absurdity and Negation

In logic we can represent the proposition ¬A by `A => Abort`. Here `Abort` is an 'absurd' proposition, *i.e.* a proposition with no conceivable proofs, or in other words an empty type.

In Aldor we can represent `Abort` by `Union()`, a union with no components, and then negation is given by `Not(A)`, which in turn is represented by `(A->Abort)`. Given a proof of absurdity, we can however prove everything; this we represent by

```
exfalso(a:Abort,B:Type):B == error "impossible value";
```

and from this we may derive a rule of contradiction

```
contraRule(B:Type,p:Not(A),q:A):B
```

which allows us to deal directly with `Not(A)`. Example proofs which involve negation, including `((A \/ ¬A)&¬¬A) => A`, can be found at

```
http://www.cs.ukc.ac.uk/people/staff/sjt/Atypical/AldorExs
```

from which all the code discussed in this paper can be downloaded.

In our discussion of absurdity we have used the `error` function to implement the law *Ex Falso Quodlibet*: from a contradiction anything can be proved. Provided that this is the *only* use of `error` in a proof then the logic given here is consistent; obviously a wider use of `error` could render the logic inconsistent.

4 Predicate Logic

This section explores the representation of predicate logic within Aldor using dependent types and the `pretend` mechanism.

How is a predicate over a type `A` to be represented? We use a *propositional function* of type `A->Type`, so that if `F` is such a predicate then `F(a)` represents the proposition that `F` holds for the value `a`.

Existential Quantification

An existential formula $(\exists x : A)F(x)$, 'there exists an `x` of type `A` for which `F(x)` holds', is given by a *record* of dependent type, and so the domain

```
Exists(A:Type,F:A->Type)
```

has the representation

```
Record(fst:A,snd:F(fst))
```

A member of this type will be a record `[a,b]` where `a` is a member of `A` and `b` is a member of `F(a)`, that is a proof of `F(a)`. This is a *constructive* interpretation of existence, since the `a` in question is an explicit *witness* to the validity of the existential statement. It is therefore easy to see that an existential proposition is introduced by a function

```
existsIntro(a:A,b:F(a)):%
```

(where recall that `%` means 'the type being defined', that is `Exists(A,F)`).

How can we use (or 'eliminate') a proof of an existential formula? We can extract either of the component parts. It is straightforward to extract the first component, which has type `A`:

```
existsElim1(p:%):A
    == (rep p).fst;
```

but extracting the second component is somewhat more involved. Suppose that `[a,b]` has type `Record(fst:A,snd:F(fst))` then `b` will have the type `F(a)`. If `p` names the whole record, then the type of its second component will necessarily involve its first, and hence the appearance of the function `existsElim1` in the type of `existsElim2`:

```
existsElim2(p:%): F(existsElim1(A,F,p))
    == (rep p).snd pretend F(existsElim1(A,F,p));
```

This definition uses the **pretend** facility by which (e pretend T) has type T irrespective of the type for e deduced by the system. This type casting mechanism is used here to compensate for the fact that Aldor does not have type evaluation. In this example it does not recognise the identity of F(fst) – the type deduced for (rep p).snd by Aldor – and F(existsElim1(A,F,p)), despite the definition of existsElim1. This is the first case where *type evaluation* plays an essential role in the definitions that we make.

Another example, a proof of (∃x:A)(P(x)&Q(x)) => (∃x:A)P(x), is given in Figure 2. In this proof **pretend** is used to identify the fact that substitution commutes with a 'lifted' form of conjunction,

```
And1(A:Type,P:A->Type,Q:A->Type):(A->Type)
    == (a:A):Type+->And(P(a),Q(a));
```

informally giving the equivalence: (P &$_1$ Q)(x) ≡ (P(x) & Q(x)).

Universal Quantification

A universal formula (∀x:A)F(x), 'for all x of type A, F(x) holds', is given by a *function* of dependent type, (x:A)->F(x). If f has this type then for each a in A, f(a) is of type F(a), that is for each a in A, f(a) is a proof of the proposition F(a). We therefore define a domain

```
All(A:Type,F:A->Type)
```

with the representation

```
existsAnd3(A:Type,P:A->Type,Q:A->Type,
          r:Exists(A,And1(A,P,Q))):Exists(A,P)
    == {
    val ==> existsElim1(r);                    -- The witness.
    import from And(P(val),Q(val));            -- To ensure that
                                               -- andElim1 can be used.
    existsIntro(val,
              andElim1(existsElim2(r)
                      pretend And(P(val),Q(val)) ) )};
```

Figure 2. A proof of (∃x:A)(P(x)&Q(x)) => (∃x:A)P(x).

```
(x:A)->F(x)
```

In many examples which use universal quantification it again becomes necessary to use `pretend` for type evaluation. In particular it is often the case that we have to identify a function application like `((x:A):B +-> e)(a)` with its result, namely `e[a/x]` (that is the expression `e` in which the argument `a` has been substituted for the parameter `x`).

Reasoning with Identity

A fundamental part of first-order reasoning is the logic of identity, under which equals can be substituted for equals: Leibnitz's principle. In order to implement this fully requires some support from the implementation, but we have implemented a limited version of equality reasoning using `pretend` and Aldor equality. The Aldor category `BasicType` contains a Boolean equality operation, and so all types supporting this interface have elements which can be compared for equality. We define the I-types which represent the proposition that two expressions are equal:

```
I(A:BasicType,a:A,b:A):Type
    == if (a=b) then Integer else Abort;
```

so that if `a` and `b` are equal any integer is a proof of this fact whereas there are no proofs that unequal elements are equal. There is no special reason for choosing to use `Integer` here: any non-empty type would do.

For reasoning with I-types we have an introduction rule that any element equals itself:

```
refl(A:BasicType,a:A):I(A,a,a)
    == trivial pretend I(A,a,a);
```

(where `trivial` is simply 0). Leibnitz's principle is embodied in the substitution rule:

```
subst(A:BasicType,a:A,b:A,eq:I(A,a,b),F:A->Type,p:F(a)):F(b)
    == p pretend F(b);
```

which states that if `F(a)` is valid (as witnessed by `p`) and if `a` and `b` are equal (as witnessed by `eq`), then `F(b)` is valid. The proof of `F(b)` is again `p`, but this time coerced into type `F(b)` by `pretend`.

As we remarked earlier, unrestricted use of `pretend` will result in an inconsistent system, but if it is encapsulated in a rule like `subst` *it can only be used when there is evidence that use of it is sound*; this is the exact role of the equality witness `eq`.

Note that this treatment is not simply syntactic sugar for the Boolean operation; we are able to perform *hypothetical* reasoning using this implementation, which is not possible using the Boolean-valued operation.

For example, we can give a general proof that identity is symmetric

```
symm(A:BasicType,a:A,b:A):(I(A,a,b) -> I(A,b,a))
```

If we are supplied with a proof of I(A,a,b) we can use that proof in an application of subst to replace by b the (boxed occurrence of) a in

I(A, a ,a)

Again, this proof will use pretend to mimic type evaluation; full proofs of symmetry and transitivity are at the Web site mentioned earlier.

This completes our discussion of an embedding of a constructive, many-sorted, logic in the type theory of Aldor, modulo use of the pretend operation. In the presence of type evaluation we will be able to eschew use of pretend.

5   Categories and Axioms

We have now developed enough logical machinery to introduce an axiomatisation of some algebraic notions. In the standard Aldor library we can find categories (that is interfaces) which are intended to capture the notions of monoid and group:

```
Monoid: Category == BasicType with {
        *          : (%,%) -> %;
        1          : %; };

Group : Category == Monoid with {
        inv        : % -> %; };
```

As we have argued elsewhere [9] these categories fail to capture what it means to be a group or a monoid since they lack any axiomatisation. We can correct that by adding axioms to the categories thus:

```
MonoidAx (M:Monoid): Category == with {
        import from M;
        leftUnit  : (m:M) -> I(M,m,1*m);
        rightUnit : (m:M) -> I(M,m,m*1);
        assoc     : (m:M,n:M,p:M) -> I(M,m*(n*p),(m*n)*p); };
```

```
GroupAx (G:Group): Category == MonoidAx(G) with {
       import from G;
       leftInv  : (g:G) -> I(G,1,g*inv(g));
       rightInv : (g:G) -> I(G,1,inv(g)*g); }
```

Note that we use the logic developed earlier to express universally quantified formulas by means of dependent functions and identity using I-types.

We are also able to use *inheritance* in writing the axiomatisation: in writing the axioms for a group G, GroupAx(G), we extend the monoid axiomatisation for the same structure, Monoid(G). In the latter expression we use the fact that Group inherits from Monoid, so that a group may be passed as a parameter at any point where a monoid is expected.

Using the techniques introduced here it is possible to build a hierarchy of *axiomatic* categories which shadows the *signature* hierarchy already implemented in Aldor. The axiomatic hierarchy allows more distinctions to be made: a commutative monoid has the same signature as a monoid but is different axiomatically.

```
CommutativeMonoidAx (M:Monoid): Category == MonoidAx(M) with {
       import from M;
       comm      : (m:M,n:M) -> I(M,m*n,n*m); };
```

In the remainder of the section we introduce the booleans as a monoid and show how they conform to the axioms.

The Logic of the `Boolean` Type

The Boolean type in Aldor consists of the values true and false. We prove that a property F – that is a propositional function F:Boolean->Type – is valid *for all* Booleans by showing that it holds for both values, that is, that we have proofs for F(true) and F(false). This proof rule is embodied in the function

```
boolAll(F:Boolean->Type,t:F(true),f:F(false),b:Boolean):F(b)
   == if b then (t pretend F(b))
           else (f pretend F(b));
```

Note that pretend is used here to give if...then...else... a dependent type. In a system with type evaluation it would be given this type, which is a generalisation of its current type.

The `Boolean` Monoid

The Booleans form an additive monoid thus:

```
(a:Boolean) * (b:Boolean) : Boolean
    == if a then (not b) else b;
1:Boolean
    == false;
```

and an illustrative proof of an axiom is given by

```
leftUnit(b:Boolean):I(Boolean,b,1*b)
 == boolAll(G,
               trivial pretend G(true),
               trivial pretend G(false),
               b) pretend I(Boolean,b,1*b)
      where {
      G:Boolean->Type == (x:Boolean):Type+->I(Boolean,x,1*x); };
```

It cannot be stressed too strongly that the three instances of **pretend** are only necessary in this proof because the current Aldor system lacks type evaluation. In such a system the undecorated term will in itself be a proof.

Other Monoids and Groups

It is possible in a similar way to show that the **Integers** form a monoid; in this case the proofs of the universal statements will be by *induction* (or equivalently, recursion). General results – such as the fact that the direct product of two monoids forms a monoid – allow other structures to be shown to be monoids. Similar considerations also apply to groups and other algebraic structures.

An Alternative Approach to Axiomatisation

The categories **MonoidAx** and **GroupAx** are parametric; for a given structure S, the axioms are expressed by the categories **MonoidAx(S)** and **GroupAx(S)**. An alternative approach is given by

```
MonoidAndAx: Category == Monoid with {
        import from %;
        leftUnit  : (m:%) -> I(%,m,1*m);
        rightUnit : (m:%) -> I(%,m,m*1);
        assoc     : (m:%,n:%,p:%) -> I(%,m*(n*p),(m*n)*p); };

GroupAndAx: Category == Join(MonoidAndAx,Group) with {
        import from %;
        leftInv  : (g:%) -> I(%,1,g*inv(g));
        rightInv : (g:%) -> I(%,1,inv(g)*g); };
```

These categories `extend` the categories `Monoid` and `Group`; the statement `import from %` ensures that the operations of the base category are visible in the extension.

Under this approach, the algebraic operations are grouped with the proofs that they satisfy the appropriate axioms. Using the `extend` operation it is possible to extend a domain with proof objects constructed with reference to the underlying representation; this is not possible under the first approach. It remains to be seen which style of axiomatisation is more flexible in practice.

6 Programming with Dependent Types: Vectors

In this section we revisit the example of vectors mentioned earlier and in [9]. As in the previous sections, we use the `pretend` operation to sidestep the typechecker, but we stress that the code written here executes perfectly well in the unmodified system, which is based on a first-order, weakly typed, abstract machine.

We have implemented a variety of operations over vectors; this section gives an overview of the code presented in Figure 3. The type of `Integer` vectors, `Vector`, is defined by recursion over the natural numbers: a vector of length 1 is simply an Integer, whilst a general vector of length n is a record [x,v] consisting of a (first) value x and a vector v of length n-1. The function `vCons` is the constructor for this type, and it is illustrated in the construction of the zero vector of length n, `zVec(n)`.

The most interesting function from the type-checking point of view is `append`, which joins two vectors of length n and m to give a vector of length n+m. The three slanted instances of *pretend* illustrate non-trivial coercions which identify n+0 and n; (m+1)-1 and m; and (n+m)-1 and (n-1)+m respectively. It is arguable how easy it is to automate these. From the point of view of a user of this library, however, when we write a concrete application like `addVec(5,append(2,3,vec2,vec3))` it is possible to recognise the type of `append(2,3,vec2,vec3)` as `Vector(5)` and thus to allow the expression to be type checked successfully.

7 Conclusions

We have shown that using the 'pretend' type casting mechanism of Aldor we are able to implement a prototype of a version of Aldor with type evaluation. This was illustrated by a number of examples.

As this is an exercise in prototyping, we have paid no attention to the consistency of the logical system; [9] discusses this issue in greater detail.

```
-- The definition of Vectors of len gth n.
Vector(n:Inte ger): Type
 == if n<=1 then Inte ger else Record(fst:Inte ger,rst:Vector(n-1));

-- Constructin g a Vector(n) from an Inte ger and a Vector(n-1).
vCons(n:Inte ger,x:Integer,v:Vector(n-1)):Vector(n)
 == { import from Record(fst:Inte ger,rst:Vector(n-1));
      [x,v]@Record(fst:Inte ger,rst:Vector(n-1)) pretend Vector(n);
};

-- A zero vector of len gth n.
zVec(n:Inte ger):Vector(n)
 == if n<=1 then 0 pretend Vector(n) else vCons(n,0,zVec(n-1));

-- Append two Vectors.
append(n:Inte ger,m:Inte ger,v:Vector(n),w:Vector(m)):Vector(n+m)
 == { if n<=0 then w pretend Vector(n+m)
      else if n=1 then vCons(m+1,
                            (v pretend Inte ger),
                            (w pretend Vector((m+1)-1)))
                      pretend Vector(n+m)
      else vCons(n+m,
              vec.fst,
              append(n-1,m,vec.rst,w) pretend
Vector((n+m)-1))
      where { vec == (v pretend
                     Record(fst:Inte ger,rst:Vector(n-1))); }};
```

Figure 3. The Vector type and functions.

The development of proofs given here shows one mechanism for inte-gration of symbolic computation and reasoning, namely that the proofs are written in the Aldor language. An alternative would be provided by interfacing the system to a theorem prover such as Coq [4], which is based on a logic similar to that reported here.

References

[1]   Andrej Bauer, Edmund Clarke, and Xudong Zhao. Analytica – an experiment in combining theorem proving and symbolic computation. In *AISMC-3*, volume 1138 of *LNCS*. Springer, 1996.

[2] Bruno Buchberger. Symbolic Computation: Computer Algebra and Logic. In F. Baader and K.U. Schulz, editors, *Frontiers of Combining Systems*. Kluwer, 1996.

[3] Jaques Calmet and Karsten Homann. Classification of communication and cooperation mechanisms for logical and symbolic computation systems. In *FroCos'96*. Kluwer, 1996.

[4] C. Cornes et al. The Coq proof assistant reference manual, version 5.10. Rapport technique RT-0177, INRIA, 1995.

[5] John Hughes and Simon Peyton Jones, editors. *Report on the Programming Language Haskell 98*. http://www.haskell.org/report/, 1999.

[6] Richard D. Jenks and Robert S. Sutor. *Axiom: The Scientific Computation System*. Springer, 1992.

[7] Per Martin-Löf. *Intuitionistic Type Theory*. Bibliopolis, Naples, 1984. Based on a set of notes taken by Giovanni Sambin of a series of lectures given in Padova, June 1980.

[8] Erik Poll and Simon Thompson. The Type System of Aldor. Technical Report 11-99, Computing Laboratory, University of Kent at Canterbury, 1999.

[9] Erik Poll and Simon Thompson. Integrating Computer Algebra and Reasoning through the Type System of Aldor In Hélène Kirchner and Christophe Ringeissen, editors, *Frontiers of Combining Systems: FroCoS 2000, LNCS 1794*. Springer-Verlag, Heidelberg, 2000.

[10] Simon Thompson. *Type Theory and Functional Programming*. Addison Wesley, 1991.

[11] Stephen M. Watt et al. A First Report on the $A^{\#}$ Compiler. In *ISSAC 94*. ACM Press, 1994.

[12] Stephen M. Watt et al. *AXIOM: Library Compiler User Guide*. NAG Ltd., 1995.

Part II

Invited Presentations

Communicating Mathematics on the Web

Henk Barendregt   Arjeh Cohen

The following claims will be made.

1. There is a language L that is enough powerful such that most mathematics can be formulated in it, enough natural that mathematicians can use it easily and enough formal that computers can deal with it. For example in L one can state that f is the primitive of the continuous function g on the reals (whereby explicit forms of f and g can be given), that α is the largest eigenvalue of the symmetric matrix M, or that $\{x_1, \ldots, x_k\}$ is the basis of the Hilbert-space H (where again the x's and H can be described explicitly).

2. There are computer systems that can answer for substantial parts of mathematics the following queries.

    (i) Does statement A hold?

    (ii) Give me an object x such that $A(x)$ holds. (Now the computer has the responsibility of giving the explicit form of x.)

    (iii) Give me evidence for the answers in (i) and (ii).

It will be argued that it is possible and desirable to construct this language L and the mentioned computer systems.

An extended version of this abstract is submitted under the title "Electronic Communication of Mathematics" to the Special Issue on "Computer Algebra and Mechanized Reasoning" of the Journal of Symbolic Computation, 2001.

Teaching Mathematics Accross the Internet

Gaston H. Gonnet

Many models have been proposed for electronic books. They all have potential, but until now, no clear winner has emerged. Based on our experience with Maple and with OpenMath (a standard to exchange mathematical data), we propose a method which makes books (heavy on mathematical content) accessible through the web via normal browsers. This is based on an architecture which allows any necessary mathematical computations to take place in computing servers. Several advantages arise from this model, the main ones being the ability to use the material from any, no matter how minimal, hardware, no installation of any particular software required and almost universal accessibility.

We show examples from various experiments, the first being a Comap book, "Principles and practice of Mathematics" where we illustrate how the material is presented and the exercises that the students can attempt. The other examples are more technical and illustrate the basis of non-linear least squares approximations, Integration and differentiation in tandem and a Hypergeometric sums calculator.

For more information, see `http://www.inf.ethz.ch/personal/gonnet/`.

Part III

System Description

SINGULAR – A Computer Algebra System for Polynomial Computations

Gert-Martin Greuel   Gerhard Pfister
Hans Schönemann

Abstract. SINGULAR *is a specialized computer algebra system for polynomial computations with emphasize on the needs of commutative algebra, algebraic geometry, and singularity theory.*

1 Main Functionality

SINGULAR's main computational objects are polynomials, ideals and modules over a large variety of rings. SINGULAR features one of the fastest and most general implementations of various algorithms for computing standard resp. Gröbner bases. Furthermore, it provides multivariate polynomial factorization, resultant, characteristic set and gcd computations, syzygy and free-resolution computations, numerical root–finding, visualisation, and many more related functionalities.

Based on an easy-to-use interactive shell and C-like programming language, SINGULAR's internal functionality is augmented and user-extendable by libraries written in the SINGULAR programming language or in C++. A general and efficient implementation of links as endpoints of communications allows SINGULAR to make its functionality available to and be easily incorporated into other programmes.

The main goal of the SINGULAR–group is to further develop and implement *advanced* algorithms to be used for mathematical research, in particular in commutative algebra, algebraic geometry and singularity theory.

Indeed there exist already several libraries providing such algorithms, including full primary decomposition for several ground fields, ring normalization (integral closure), versal deformations of arbitrary isolated singularities, monodromy and spectral numbers for hypersurface singularities, Hamburger–Noether (Puiseux)–expansions of plane curve singularities and many more. Most of these algorithms are not available in any other system.

Recently, due to nonmathematical applications of SINGULAR, we are experimenting with symbolic–numerical polynomial solving.

As a specialized system, SINGULAR's aim is not to provide all the functionality of a general purpose system. The main strength of the system, besides the above mentioned functionality, is the speed of the important basic algorithms such as Gröbner basis, syzygy, and free resolution computations for modules. It is impossible to detail any of the algorithms, we rather refer to the literature, given in the references.

SINGULAR's online help system is available in various formats where the HTML format is especially user-friendly.

2　Availability

SINGULAR is publicly available as a binary program for all common Unix platforms including LINUX, for Windows 95/98/NT and for MacOS.

The current version number is 1.3.8. It can be downloaded by anonymous ftp from `ftp://www.mathematik.uni-kl.de/pub/Math/Singular` or per WWW from `http://www.singular.uni-kl.de/distribution.html`

Besides the executable SINGULAR program, the distribution contains the source code of all SINGULAR libraries, the user manual (resp. tutorial) in various formats (PostScript, info, and HTML) and some utility programs (for visualisation etc.).

Moreover, on request, we provide the source code of the FACTORY library for multivariate polynomial gcd, resultant, and factorization which is part of SINGULAR but may be used by other systems (as it is e.g., by Macaulay 2).

For more and always up-to-date information, SINGULAR's home page can be reached at `http://www.singular.uni-kl.de`

3　Mathematical Features

SINGULAR's primary computational objects are ideals resp. modules which are generated by polynomials resp. polynomial-vectors over polynomial rings or, more generally, over the localization of a polynomial ring with respect to any ordering on the set of monomials which is compatible with the semigroup structure.

Supported baserings include:
- polynomial rings with a large variety of polynomial orderings (common simple, block, elimination, weighted, and general matrix orderings),
- localization of a polynomial ring at a prime ideal generated by a subset of the variables
- factor rings by an ideal of one of the above,
- rings of tensor products of one of the above.

Supported ground fields for these rings include:
- rational numbers (of arbitrary length) Q,
- finite fields Z/p (where p is a prime ≤ 32003),
- Galois fields (finite fields with $q = p^n \leq 2^{15}$ elements),
- transcendental extensions $(K(A, B, C, \dots), K = Q$ or $Z/p)$,
- algebraic extensions $(K[t]/\text{minimal-polynomial}, K = Q$ or $Z/p)$, and
- floating point real and complex numbers with arbitrary predefined precision.

The main algorithms implemented in SINGULAR are:
- general standard basis algorithm for *any* monomial ordering which is compatible with the natural semi-group structure of the exponents. This includes well-orderings (Buchberger algorithm) and tangent cone orderings (Mora algorithm) as special cases,
- Hilbert–driven Gröbner basis algorithms, weighted–ecart–method and high–corner–method, FGLM algorithm for change of ordering,
- factorizing Buchberger algorithm,
- intersection, quotient, elimination and saturation of ideals,
- Schreyer's, La Scala's and Siebert's algorithm for computations of syzygies and free resolutions of modules,
- combinatorial algorithms for computations of dimensions of factor rings, Hilbert series and multiplicities of modules,
- multivariate polynomial gcd, resultant, and factorization algorithms,
- Wang's algorithm to compute characteristic sets.

See [2, 3, 5, 4] for more details on the implemented algorithms.

4 Computational Features

SINGULAR has a convenient and intuitive interactive user interface (shell) which has key-bindings similar to those of Unix' `tcsh` shell. Alternatively, an Emacs mode provides the possibility to run SINGULAR within an Emacs/XEmacs window. SINGULAR's user interface provides both, access to SINGULAR's mathematical functionality and a convenient, powerful, and C-like programming language (strongly typed, and lexicographically scoped) which includes all the usual programming constructs (like loops, procedures, local/global variables, etc). Based on this programming language, users may extend SINGULAR's functionality by writing their own libraries. Libraries may be written in the SINGULAR programming language or in C++.

At the moment, the SINGULAR distribution includes the following libraries:

standard.lib   extensions of Singular kernel

General purpose
all.lib   load all other libraries
general.lib   procedures of general type
inout.lib   procedures for manipulating in- and output
poly.lib   procedures for manipulating polynomials and ideals
random.lib   procedures of random/sparse matrix and poly operations
ring.lib   procedures for manipulating rings and maps

Linear algebra
matrix.lib   procedures for matrix operations
jordan.lib   procedures to compute the jordan normal form
linalg.lib   procedures for algorithmic linear algebra

Commutative algebra
algebra.lib   procedures for computing with algebras and maps
elim.lib   procedures for elimination, saturation and blowing up
homolog.lib   procedures for homological algebra
mregular.lib   procedures for Castelnuovo-Mumford regularity
normal.lib   procedures for normalization
primdec.lib   procedures for primary decomposition
primitiv.lib   procedures for finding a primitive element
intprog.lib   procedures for integer programming
toric.lib   procedures for toric ideals

Singularities
classify.lib   procedures for the Arnold-classifier of singularities
deform.lib   procedures for computing miniversal deformation
hnoether.lib   procedures for Hamburger-Noether (Puiseux) development
mondromy.lib   procedures to compute the monodromy of a singularity
sing.lib   procedures for computing invariants of singularities
spcurve.lib   procedures for Cohen–Macaulay codimension 2 singularities

Invariant theory

finvar.lib	procedures to calculate invariant rings of finite groups
ainvar.lib	procedures for invariant rings of the additive group

Symbolic-numerical solving

presolve.lib	procedures for pre-solving polynomial equations
solve.lib	procedures to solve polynomial systems
triang.lib	procedures for decomposing zero-dimensional ideals
ntsolve.lib	procedures for real Newton solving

Visualization

graphics.lib	procedures to draw with Mathematica
latex.lib	procedures for typesetting in TeX
surf.lib	interface to the surf programme

5   Links

SINGULAR furthermore features links as general endpoints of communications, i.e. as something, SINGULAR can read from or write to. To this point, the following link types are implemented:

Ascii text	Output can conveniently be viewed and manipulated. Read/write is not the fastest.
DBM	Provides access to data stored in a data base.
MP file	Stores data in the binary Multi Protocol (MP) format [1]. Read/write is very fast.
MP TCP	Exchanges data in binary MP format between processes (on the same or different computers); data exchange is very efficient

The functionality of theses links is provided to the user by a general, consistent and convenient link interface.

Based on MP TCP links, SINGULAR can very efficiently communicate with itself which opens the door for implementations of parallel/distributed algorithms. Furthermore, the same links can be used to communicate with other Computer Algebra programs which have an MP interface. At the moment, there are MP interfaces for MuPAD and Mathematica, enabling the use of one system from the others (e.g. one can use SINGULAR's functionality from within Mathematica or MuPAD).

6   Performance

SINGULAR's kernel is implemented in C/C++. SINGULAR's main implementation design goal is speed of the mainly used algorithms. Therefore, all time-consuming operations like standard bases computations or factorization are implemented in its kernel. As another consequence, SINGULAR has the concept of a global ring which needs to be defined prior to any polynomial operations. Arbitrary precision integer and floating point arithmetic is accomplished by linking SINGULAR with the GNU multiple precision library gmp and modulo arithmetic is accomplished by using look-up tables. Polynomials are internally represented as linked lists of monomials, where a monomial consists of a coefficient and an exponent vector.

To illustrate SINGULAR's computing speed, the table below shows timings of various systems for solving problem 6 of the ISSAC'97 system challenge (computation of a lexicographical Gröbner basis, 16 roots). The first row shows the timings for the Gröbner basis computation (using the default command as the respective manual describes them). The second row shows the time for numerical solving (finding all complex roots) of the same system (commands NSolve (Mathematica), evalf(solve(..)) (Maple) and triangL_solve (Singular)). (The new symbolic-numerical algorithms for polynomial system solving in SINGULAR seem to be especially promising, cf. [6]). All timings were taken on a Pentium Pro 200 with 128 MB of RAM running Linux:

	Mathematica 4.0	Maple V.5.1	GB v3	SINGULAR
Gröbner basis	159.49 sec	14.66 sec	207 sec	0.61 sec
numerical solution	15.91 sec	out of mem.	-	4.28 sec

Computations with MuPAD 1.4, Reduce, CoCoA (3.0.2) and Macaulay 2 (0.8.14) could not finish the Gröbner basis computation within 15 CPU hours.

For more examples see: http://www.symbolicadata.org, a project benchmarking CA system.

7 SINGULAR 2.0

The new version, to be released at the end of 2000, has a new internal engine for polynomial arithmetic which dramatically increases its computational efficiency while at the same time decreases its memory usage.

ESINGULAR is a new program for out–of–the–box, precustomized Emacs/ XEmacs which runs SINGULAR. Hence, using ESINGULAR, such features as colour input–output highlighting, pull–down menus, line truncation, a demo mode and tab completion are provided. SINGULAR 2.0 has also a new Windows port which includes ESINGULAR.

8 Future Work

SINGULAR's development is an actively ongoing project. Currently, the following features are under development:

- more flexible ring concepts
- faster computations by better adaption of datastructures to algorithms
- Newton polyhedron algorithms
- integral closure of an ideal
- primary decomposition for modules

Furthermore, an independent version SINGULAR::PLURAL is under development which is able to compute standard bases and syzygies in very general *non–commutative* structures with applications especially for Lie algebras.

References

[1] O. Bachman, S. Gray, and H. Schönemann. MP: A Frame work for Distributed Polynomial Systems Based on MP. In *Proc. of the International Symposium on Symbolic and Algebraic Computation (ISSAC'96)*, Zurich, Switzerland, July 1996, ACM Press.

[2] H. Grassmann, G.-M. Greuel, B. Martin, W. Neumann, G. Pfister, W. Pohl, H. Schönemann, and T. Siebert. On an implementaion of standard bases and syzygies in SINGULAR. *Computational methods in Lie Theory*. AAECC, 7:235-249, 1996.

[3] G.-M. Greuel and G. Pfister. Advances and improvements in the theory of standard bases and syzygies. *Arch. d. Math.*, 66:163-176, 1996

[4] G.-M. Greuel and G. Pfister. Groebner bases and algebraic geometry. In B. Buchberger and F. Winkler, eds., *Groebner bases and applications*, volume 251 of *London Mathematical Society Lecture Notes*, pages 109-143, Cambridge University Press, 1998

[5] G.-M. Greuel, G. Pfister, and H. Schönemann. Singular version 1.2 User Manual. In *Reports On Computer Algebra*, number 21, Center for Computer Algebra. University of Kaiserslautern. June 1998.

[6] G.-M. Greuel. Applications of Computer Algebra to Algebraic Geometry, Singularity Theory and Symbolic-Numerical Solving. To appear in *Proceedings of the 3rd ECM, Barcelona 2000*

Part IV

Posters

Integration of Automated Reasoners: a Progress Report

Alessandro Armando   Silvio Ranise   Daniele Zini

The integration of automated reasoners is recognized as one of the most promising approaches to the construction of a new and more powerful generation of mechanized reasoning systems (either automated theorem provers, computer algebra systems, or model checkers). However, as witnessed by some pioneering attempts, several difficulties must be tackled in order to build compound reasoning systems which are more than the simple sum of their components:

Modeling. A fundamental problem in interfacing automated reasoners is to devise an integration schema ensuring some basic properties of the compound system such as soundness, completeness, and efficiency.

Specification. A related problem is that of specifying the integration schema and the functionalities provided by the component systems.

Support to Systems Integration. The problem here is that of achieving interoperability among automated reasoners.

The Mechanized Reasoning Group at the University of Genova has been working on the above problems since 1991 and made the following contributions:

Modeling: Integration of Decision Procedures. In [3] the first two authors have rationally reconstructed the integration schema used by Boyer and Moore to incorporate a decision procedure into their prover [6] and lifted it to a distributed version which allows for the use of a stand alone decision procedure. In [4] the first two authors have proposed a generalization of Boyer and Moore's integration schema, called Constraint Contextual Rewriting (CCR for short). Soundness and termination of CCR have been formally proved. We have also built an automated theorem prover based on CCR called **RDL** (the system distribution can be accessed via the *Constraint Contextual Rewriting Home Page* at the URL http://www.mrg.dist.unige.it/ccr).

237

Specification: Open Mechanized Reasoning Systems (OMRS).
We have contributed to the definition of the control layer of the OMRS
specification framework [1]. An OMRS specification (which is layered in
a logic, control, and interaction components) allows for an additional and
complementary way to structure the specifications w.r.t. the standard software engineering approach based on modularity. This domain-specific feature of the OMRS specification framework is fundamental to cope with the
complexity of functionalities provided by state-of-the-art implementations.

Support to Systems Integration: The Logic Broker Architecture.
In [5] we have presented the Logic Broker Architecture, a framework which
provides the needed infrastructure for making mechanized reasoning systems interoperate. The architecture provides location transparency, a way
to forward requests for logical services to appropriate reasoning systems
via a simple registration/subscription mechanism, and a translation mechanism which ensures the transparent exchange of logical services.

Future Work. We plan to define the interaction layer of OMRS ([2]
reports a first exploratory step in this direction) and to lift the Logic Broker
Architecture to an agent-oriented architecture.

References

[1] A. Armando, A. Coglio, and F. Giunchiglia. The Control Component of
Open Mechanized Reasoning Systems. *Electronic Notes in Theoretical
Computer Science*, 23(3):3–20, 1999.

[2] A. Armando, M. Kohlhase, and S. Ranise. Protocols for Mathematical
Services Based on KQML and OMRS. In *Proc. of the 8th Symp. on the
Integ. of Symb. Comp. and Mech. Reas.*, St. Andrews, Aug. 2000.

[3] A. Armando and S. Ranise. From Integrated Reasoning Specialists
to "Plug-and-Play" Reasoning Components. In AISC'98, LNCS 1476,
pages 42–54, Plattsburgh (USA), 1998.

[4] A. Armando and S. Ranise. Termination of Constraint Contextual
Rewriting. In *Proc. of 3rd Intl. Workshop on Frontiers of Combining
Systems*, LNAI 1794, pages 47–61, Nancy, France, March 2000.

[5] A. Armando and D. Zini. Towards Interoperable Mechanized Reasoning
Systems: the Logic Broker Architecture. In *Proc. of the AI*IA-TABOO
Joint Workshop 'Dagli Oggetti agli Agenti,'* Parma, Italy, May 2000.

[6] R.S. Boyer and J S. Moore. Integrating Decision Procedures into Heuristic Theorem Provers: A Case Study of Linear Arithmetic. *Machine
Intelligence*, 11:83–124, 1988.

Algorithmic Theories and Context

Clemens Ballarin   Jacques Calmet

Abstract. *Integration of computer algebra and automated deduction aims at a more uniform way to structure mathematical knowledge. This is traditionally either represented explicitly, as stored facts in a theorem prover, or implicitly, in the algorithms of a computer algebra system.*

The tight coupling of algorithms and their specifications leads to algorithmic theories. *One way of obtaining such a coupling is through* schemata. *The concept of schemata enables a meaningful definition of the* context of *a computation. Contexts in turn make the coupling memory-efficient.*

1   Schemata

Schemata present algorithms together with their specification in a uniform way.[1] A schema essentially extends an algorithm by its *specification*, and it makes *sub-algorithms* explicit.

In a layer view of symbolic computation systems,[2] data and theorems are the objects that are manipulated by algebraic algorithms and deduction rules. Built on top of this, the *control layer* contains knowledge on how to use these objects in a meaningful way — for example, heuristics to guide proof search or to solve a computational problem.

We propose to embed algorithmic theories based upon schemata into this layered architecture. In this framework, the execution sequence is not only determined by the algorithms but also by the control, which makes use of the specifications. During the execution some reasoning involving the

---

[1]Homann, K. and Calmet, J. (1995). Combining theorem proving and symbolic mathematical computing. In J. Calmet and J. A. Campbell, *Integrating Symbolic Mathematical Computation and Artificial Intelligence*, LNCS 958, pages 18–29.

[2]Bertoli, P. G., Calmet, J., Giunchiglia, F., and Homann, K. (1999). Specification and integration of theorem provers and computer algebra systems. *Fundamenta Informaticae*, **39**(1,2), 39–57.

specification of the algorithms will take place. Consider, for example, the simplification process in a computer algebra system. Here the control ought to keep track of divisions by expressions such as $x - a$ to avoid possible divisions by zero later on in a computation.

2 Context

In order to do so, a *context* of the computation needs to be maintained at the control level. This context has a structure similar to the execution-stack of the computation, but it contains more information.

The context contains references to the schemata whose algorithms are used. This part is static. It also contains instantiations of constraints and specifications by arguments or results of computations. In the simplification example, equations of the form $x \neq a$ would be added. This part of the context is dynamic.

The dynamic part of the context grows and shrinks during a computation. It is related to the execution of the algorithm and can be discarded after its termination, and after its result has been assessed by the control. Consequently, not the whole history of a computation needs to be stored. For example, only the operations that could lead to mathematically unsound operations need be kept track of by the control and stored until soundness can be established.

3 Conclusion

We have shown how to use context to deal with the dynamics in an integration of computation and deduction. By avoiding to store the entire history of a computation such an integration becomes feasible for lengthy computations.

Tighter coupling of both computation and deduction makes the software design process less rigid. Traditionally, an algorithm is designed together with its specification and then implemented. Any unintentional use would require one to redesign the algorithm and then rewrite or modify the code. By giving the specification the same status as the code, it is possible to inspect both and to reason about them. For example, during a computation it might be possible to establish correctness even if the domain of computation does not meet the specification. Another application is to use an implementation that is not trusted fully and to establish correctness explicitly.

The GiNaC Framework for
Symbolic Computation within the
C++ Programming Language

Christian Bauer   Alexander Frink   Richard Kreckel

Abstract. *GiNaC is an Open Source C++ library for symbolic computation. It was originally designed for applications in high-energy physics but it can be used as a general symbolic computation engine, especially for programs that combine symbolic computation with numerical methods, graphical user interfaces, deduction, networking etc. GiNaC was designed to overcome the problems arising from the multi-lingual situation one usually encounters with traditional Computer Algebra Systems.*

1 Overview

When we start a software project that relies to some extent on manipulating symbolic expressions (as opposed to quickly checking some result with our favoured CAS), we are usually faced with a multi-lingual situation. We start by implementing some formulae using a symbolic package and the language it provides. Then we want to get numerical results out of it which is usually done by the CAS' code-generator which produces C or FORTRAN code which we compile and let run. Sometimes, we also wish to provide an intuitive user interface for our program.

It is not uncommon to see how the interaction of the different software packages in use makes the whole endeavor fail. It may be that our symbolic language is too restriced. It may also be that our programs or the scripts we wrote to glue everything together break at each software upgrade. It may even turn out that we cannot convince our colleagues to help us with coding in three different languages on one single project. This is a situation not uncommon in real-world projects which GiNaC was designed to overcome.

2 An Example

GiNaC deliberately denies the need for a distinction of implementation language at different steps of a project. It is entirely written in C++ and adheres to the ISO standard for C++. The user can interact with it directly in that language, freely build upon it and extend it (try extending Mathematica's Kernel!). Here is a complete program that computes Hermite polynomials:

```
1  #include <ginac/ginac.h>
2  using namespace GiNaC;
3
4  int main(int argc, char **argv)
5  {
6      int degree = atoi(argv[1]);
7      numeric value = numeric(argv[2]);
8      symbol z("z");
9      const ex HGen = exp(-pow(z, 2));
10     ex H = normal(pow(-1, degree) * HGen.diff(z, degree) / HGen);
11     cout << "H_" << degree << "(z) == " << H << endl;
12     cout << "H_" << degree << "(" << value << ") == "
13         << H.subs(z == value) << endl;
14     return 0;
15 }
```

Syntactically, the program shows how symbolic expressions are written down in GiNaC pretty much like common numeric terms, thanks to operator overloading.

3 The Implementation

GiNaC implements a number of symbolic classes, all of which are referenced by the class of all expressions **ex**. There is reference counting at work, implemented in the interplay between class **ex** and the abstract base class **basic**, so the user who wishes to extend the system does not have to worry about memory management. For numeric types, GiNaC uses Bruno Haible's super-efficient library CLN (`http://clisp.cons.org/~haible/packages-cln.html`). Numerous interactive interfaces to GiNaC are imaginable. The most intriguing one is an interface to Masaharu Goto's C++-interpreter Cint (`http://root.cern.ch/root/Cint.html`).

4 Status and Availability

Being a special-purpose system, GiNaC aims at being a fast and reliable foundation for combined symbolical/numerical/graphical projects in C++. It may be downloaded and distributed under the terms of the GNU General Public License from `http://www.ginac.de/`. A tutorial introduction and complete cross references of the source code can also be found there.

A Framework for Propositional Model Elimination Algorithms

Marco Benedetti

The satisfiability problem of propositional formulas by means of model elimination is a field where the proposals of algorithms have been abundant. We propose a unifying framework to enlighten and capture some hidden schemata shared by different enhancements to the basic model elimination procedure[1]. Among these enhancements we recall the use of lemmas to memorize successful sub-refutation attempts, the use of mechanisms to memorize failures of sub-refutation attempts, the addition of constraint-propagation and forward-checking techniques and the introduction of heuristics able to drive the search in a suitable way. The key idea behind our approach is to distinguish the core search procedure of the ME method from every superimposed strategy used to drive or prune the search. The distinction is achieved at the design stage with the introduction of an *adviser* and the consequent definition of a coupled *model elimination – adviser* architecture called ME-A. *Advices* coming from the *adviser* cause the search procedure to (hopefully) behave in a smart and non redundant way. Mechanisms needed to give *good* advices are entirely hidden in the *adviser*, while the basic ME search procedure is just slightly revised to allow interaction with the *adviser*. This means that the basic ME search procedure – once properly connected to the adviser – remains untouched, despite of the large number (and different kinds) of refinements and enhancements that can be applied. So, the main concern of the coupled ME-A architecture is to be general enough to capture in a fixed and simple schema as many enhancements to ME as possible.

To achieve such result, a distinction between *logical* and *extra-logical* issues is made, and the adviser is accordingly specialized in a logical section (called $adviser^L$) and an extra-logical section (called $adviser^{EL}$). The point about this distinction is that every piece of information that makes any

---

[1] A more detailed description of our architecture can be found at
ftp://www.dis.uniroma1.it/~benedett/MEA/mea.ps

sense with respect to the formula alone is regarded to be *logical*, whereas those procedural aspects that are a contingency with respect to the meaning of a SAT problem – being more closely related to the search control – are regarded as *extra-logical* facts.

The logical section of the adviser receives and handles a growing bundle of logical knowledge, arising from sub-refutation outcomes, and stores it for later reuse. Conversely, the search procedure receives an advice every time it is possible for the adviser to foresee the outcome of the sub-refutation attempt which is going to be performed. This way the search space is heavily pruned, embodying the reasonable behaviour to avoid sub-search whose partial result is (directly or not) already known.

The bridge between the *ME* procedure and the adviser allows for passing extra-logical knowledge too. This time, the mechanism we use is to empower the $adviser^{EL}$ to give advices about every choice arising during the search, so to make it responsible for actually choosing the search strategy, i.e. for finding a good path through the non-deterministic structure of the refutation mechanism.

Finally, coordination between $adviser^L$ and $adviser^{EL}$ can be exploited to achieve an interesting behaviour of the *ME-A* algorithm when the logic point of view conflicts with the procedural limitation of the basic algorithm. For example, this cooperation can be used as a workaround to avoid to ever *chronologically* backtrack on subgoals.

We implemented a propositional SAT solver based upon the principles discussed above, and we observed three kinds of benefits resulting from experimenting within the *ME-A* framework. First of all, many known improvements to *ME* can be viewed as particular *ME-A* algorithms, provided certain properties of the adviser (discussed elsewhere) are guaranteed. In this sense, the modularity and generality of our architecture constitute a novelty which plays a great role in achieving a better understanding of existing *ME* variants. Strictly related to the possibility of uniformly describing *ME* variants is the opportunity of testing and comparing them on a fixed implementation environment, thus giving us the chance to experimentally compare different approaches with reliable results. Last but not least, a natural step from better-understanding to better-developing has been taken, building advisers (described elsewhere) that use more efficient search-pruning strategies.

Resource Guided Concurrent Deduction*

C. Benzmüller   M. Jamnik   M. Kerber   V. Sorge

Motivation Our architecture is motivated by some findings in cognitive science and particularly reflects Hadamard's "Psychology of Invention". It builds on Bundy's proof planning idea.

System Architecture The architecture (for further details see Technical Report CSRP-99-17, University of Birmingham, School of Computer Science, 1999) that we describe here allows a number of proof search attempts to be executed in parallel. Each specialised subsystem may try a different proof strategy to find the proof of a conjecture. Hence, a number of different proof strategies are used at the same time in the proof search. However, following all the available strategies simultaneously would quickly consume the available system resources consisting of computation time and memory space. In order to prevent this, and furthermore, to guide the proof search, we developed and employ a resource management concept in proof search. Resource management is a technique which distributes the available resources amongst the available subsystems. Periodically, it assesses the state of the proof search process, evaluates the progress, chooses a promising direction for further search and redistributes the available resources accordingly. If the current search direction becomes increasingly less promising then backtracking to the previous points in the search space is possible. Hence, only successful or promising proof attempts are allowed to continue searching for a proof. This process is repeated until a proof is found, or some other terminating condition is reached. An important aspect of our architecture is that in each evaluation phase the global proof state is updated, that is, promising partial proofs and especially solved subproblems are reported to a special plan server that maintains the progress of the overall proof search attempt. Furthermore, interesting results may be communicated between the subsystems (for instance, an open subproblem may be passed to a theorem prover that seems to be more appropriate).

---

*This work is partly supported by EPSRC grant GR/M99644.

245

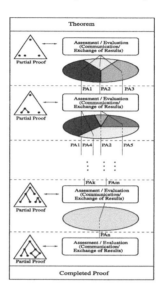

Figure 1. Distribution of Resources.

This communication is supported by the shells implemented around the specialised problem solvers. The resource management mechanism analyses the theorem and decides which subsystems, i.e., which provers, should be launched and what proportion of the resources needs to be assigned to a particular prover. The mechanism is also responsible for restricting the amount of information exchange between subsystems, so that not all of the resources are allocated to the communication. Figure 1 demonstrates this concurrent resource management based proof planning architecture. The involved planning agents are represented by PA_n and the ovals indicate the amount of resources assigned to them in each reasoning phase.

We argue that the effect of resource management leads to a less brittle search technique which we call focused search. Breadth-first search is robust in the sense that it is impossible to miss a solution. However, it is normally prohibitively expensive. Heuristic search may be considered as the other extreme case: it is possible to go with modest resources very deep in a search tree. However, the search is brittle in that a single wrong decision may make it go astray and miss a solution, independently of how big the allocated resources are. Focused search can be considered as a compromise — it requires more resources than heuristic search, but not as much as breadth-first search. As a result, a solution can still be found even if the focus of the search is misplaced. Clearly, more resources are necessary in the case of a bad than of a good focus.

Automated 'Plugging and Chugging'

Simon Colton

1 Introduction

In [1], Paul Zeitz discusses how to solve mathematical problems. One technique he proposes is to calculate some examples of objects in the problem statement and try to spot a pattern or property which provides insight into the nature of the problem. Zeitz suggests:

> Plug in lots of numbers. Keep playing around until you see a pattern ... and try to figure out why the pattern you see is happening. It is a well kept secret that much high-level mathematical research is the result of low-tech "plug and chug" methods.

Zeitz gives an example taken from a Hungarian mathematics contest: prove that the product of four consecutive natural numbers cannot be the square of an integer. We have applied the HR program [2] to produce suggestions about this problem which may give the user an insight into its solution.

2 Theory Formation to Produce Suggestions

In number theory, starting with only a few background concepts such as multiplication, HR can form thousands of concepts such as prime numbers, etc. For the above problem we plugged $n = 1, 2, 3$ and 4 into the formula for the product of four consecutive integers: $n(n+1)(n+2)(n+3)$. The results from the calculation were: $24, 120, 360$ and 840. We then asked HR to form concepts in number theory and output the definition for any number type which the numbers $24, 120, 360$ and 840 all satisfied. The first answer it produced was: even numbers, which didn't help with the problem.

After a few minutes, HR's 15th suggestion was that the above numbers are all squares minus one. This is the eureka step that Paul Zeitz was hoping the reader would make through the plug and chug technique. To finish the solution, the user must show that the product of 4 consecutive integers is always a square minus one, using some algebraic manipulation:

$$n(n+1)(n+2)(n+3) = (n^2 + 3n + 1)^2 - 1$$

Finally, they must state that the distribution of the positive square numbers means that a square minus one is never a square, hence the product of four consecutive integers is never a square.

3   Conclusion

We have detailed one way in which calculation can be used to help deduction. One can imagine the initial calculations and algebraic manipulation being undertaken by a computer algebra system and the transformed conjecture being easier for an automated theorem prover to solve (although this is possibly not the case for the above example). However, the main application of this approach may not be to produce entire proofs of theorems, but to make intelligent suggestions to the user about tricky problems. Transforming the conjecture involved an inventive step whereby a new concept was introduced. We have shown how this can be done by a theory formation program such as HR, although other methods involving knowledge bases and/or machine learning techniques may be faster.

To conclude his discussion about the Hungarian problem, Zeitz says:

> Getting to the conjecture was the crux move. At this point the problem metamorphosed into an exercise!

This suggests that invention should not be overlooked when combining calculation and deduction to improve automated mathematics.

Acknowledgments

This work is supported by EPSRC grant GR/M98012.

References

[1] P. Zeitz. *The Art and Craft of Problem Solving*. John Wiley and Sons, 1999.

[2] S. Colton, A. Bundy, and T. Walsh. HR: Automatic concept formation in pure mathematics. In *Proceedings of the 16th International Joint Conference on Artificial Intelligence*, 1999.

Integrating SAT Solvers
with Domain-specific Reasoners

Fausto Giunchiglia   Roberto Sebastiani
Paolo Traverso

In the last years we have witnessed an impressive advance in the efficiency of
SAT techniques, which has brought large previously untractable problems
at the reach of state-of-the-art solvers. As a consequence, some hard real-
world problems have been successfully solved by encoding them into SAT.
Propositional planning [5] and boolean model-checking [2] are among the
best achievements.

Unfortunately, simple boolean expressions are not expressive enough
for representing many real-world problems. For example, problem domains
like resource planning, temporal reasoning, verification and model checking
of analog circuits require also handling constraints on real-valued quantities. Moreover, some problem domains like model checking often require an
explicit representation of integers and arithmetic operators, which simple
boolean expressions are not good at representing efficiently.

In 1996 we have proposed a new general approach, called SAT based, for
building domain-specific decision procedures on top of SAT solvers [4, 3].
We assume that the input is a boolean combination of domain-specific expressions –which cannot be further decomposed propositionally. We call
the latter domain-specific *atoms*. Examples of such atoms are boxed formulas like $\Box\phi$ and inequalities like $(x_i \leq y_i + c_i)$. The basic idea is to
decompose the search into two orthogonal components, one purely propositional component and one purely domain-specific component, so that to
use a SAT solver for the former and a pure domain-specific procedure for
the latter.

The SAT based approach is based on a generate-and-check paradigm.
The input formula is encoded as a propositional formula on both boolean
and domain-specific atoms, and a SAT solver is run over it. Each time an
assignment is found which propositionally satisfies the input, its consistency
is verified wrt. the domain-specific semantics of the non-boolean atoms by
a proper procedure; this is done until either one domain-consistent assign-

249

ment is found –i.e., the input formula is consistent– or no more assignments are available –i.e., the input formula is inconsistent.

So far the SAT based approach proved very effective in various problem domains like, e.g., modal logics [3], temporal reasoning [1], resource planning [6]. In particular, the latter two proved rather efficient in combining SAT reasoning with Linear Programming (LP). Now we want to focus our research on combining SAT non only with LP systems, but also with math constraint solvers and/or computer algebra systems.

The research directions are manifold. First, in our approach the SAT solver is no more used to find one single assignment, but rather to generate up to an exhaustive set of assignments. Thus, new techniques and heuristic ad hoc for this new task have to be studied. Second, to maximize the effects of techniques like backjumping and forward checking, the SAT solver has to get as much information as it can from any failed assignment. Thus, the research should be directed on maximizing the effectiveness of the information exchange between the SAT solver and the math constraint solver. Finally, the math constraint solver should not be repeatedly run on similar subproblems. This requires studying techniques for incremental constraint solving and to maximize the integration of the two solvers.

References

[1] A. Armando, C. Castellini, and E. Giunchiglia. SAT-based procedures for temporal reasoning. In *Proc. ECP-99*, 1999.

[2] A. Biere, A. Cimatti, E. Clarke, and Y. Zhu. Symbolic model checking without BDDs. In *Proc. CAV'99*, 1999.

[3] F. Giunchiglia and R. Sebastiani. Building decision procedures for modal logics from propositional decision procedures - the case study of modal K(m). Technical Report 9611-06, IRST, 1996. To appear on *Information and Computation*, Vol. 160, 2000.

[4] F. Giunchiglia and R. Sebastiani. A SAT-based decision procedure for ALC. In *Proc. KR'96*, 1996.

[5] H. Kautz, D. McAllester, and Bart Selman. Encoding Plans in Propositional Logic. In *Proc. KR'96*, 1996.

[6] S. Wolfman and D. Weld. The LPSAT Engine and its Application to Resource Planning. In *Proc. IJCAI'99*, 1999.

Solving Integrals at the Method Level[*]

Alex Heneveld   Ewen Maclean   Alan Bundy
Jacques Fleuriot   Alan Smaill

1   Modelling What People Do

We are using calculus to study proof-planning at the method level. This is a particularly well-suited domain because people tend to use higher-level methods for calculus, whilst they can when needed generate more rigourous analysis proofs for each method step.[1] We have identified the principal methods people use for integration and differentiation from textbooks and work transcripts, and we have implemented many of them in $\lambda Clam$ (Richardson, http://dream.dai.ed.ac.uk/software/systems/), a proof-planner operating on the level of methods which outputs a plan suggesting tactics a theorem prover can use to prove each method step.

Alex Heneveld is investigating how people choose an integration method from the set of possibilities. Two major rival psychological theories under review are the precondition/production-rule approach (Newell & Simon, *Human Problem Solving*, 1972; Anderson's ACT-R architecture, http://act.psy.cmu.edu; and $\lambda Clam$) and featural reminding (Gentner & Forbus's MAC/FAC theory, http://www.qrg.nwu.edu/ideas/macfac.htm; and indexes in case-based reasoning). Models for these theories are being fleshed out on the basis of experiments with students at the University of Edinburgh, and implemented and tested in $\lambda Clam$. The best method selection mechanisms for calculus may lead to better ways to teach humans to solve integrals, and a comparison of these mechanisms will offer more

[*]We acknowledge Julian Richardson, Jeremy Gow, and Louise Dennis for their assistance with $\lambda Clam$, and the Marshall Aid Commemmoration Commission and an EPSRC research grant for funding this research.

[1]Research into automated integration was popular in the 1960's (notably, SIN and SAINT used methods), but was "solved" by the discovery of the Risch-Norman algorithm. However, for our interests — cognitive modelling of method selection and automated proofs of these methods — it is still very relevant.

insight into the choices available for both automated theorem proving and proof-planning at the method level.

2   Non-standard Analysis Proofs of Calculus Methods

A major attraction of using a method-level planner in this domain is that each integration method is formally correct; once the plan is found using methods, a theorem prover can develop a formal, low-level analysis proof. Ewen Maclean is developing automated proofs for important limit and calculus theorems using the intuitive domain of non-standard analysis.[2] $\lambda Clam$ can so far automatically generate formal proofs for the chain rule and integration by parts (the product rule) over the hyperreals, and other theorems and methods will follow soon. However, one major outstanding issue remains: we would like $\lambda Clam$ to prove the transfer of validity results from the hyperreals to standard analysis fields (R and C).

3   Symbiosis with Computer Algebra

Some of the methods to solve integrals and some analysis proofs have been tricky to automate due to the weak algebra facilities in $\lambda Clam$. One promising option is to call a computer algebra system from $\lambda Clam$: this should be in place soon, and meanwhile we welcome advice on this topic.

We believe that such a link will have reciprocal benefits for computer algebra packages. In the near term, there are three obvious benefits: firstly, CA systems can use $\lambda Clam$ to solve problems in non-standard analysis; secondly, CA systems can perform individual integration steps at the method level by calling $\lambda Clam$'s methods; and thirdly, CA systems can completely solve integrals using techniques the user recognises. Eventually, we envision computer mathematics tools typically interacting with the user on the method level, in all mathematics domains, where methods are problem solving steps the user is accustomed to working with. The particular applications of our work to this eventual CA/proof-planning marriage will be in (1) suggesting good methods to apply to a problem, and (2) supplying a low-level, formal proof for a method-level solution plan.

---

[2]Non-standard analysis (Robinson, *Non-standard Analysis*, 1966) defines the infinitesimal (ϵ, where $\epsilon > 0$ is a quantity smaller than all positive reals) and is performed over the hyperreals (the field of reals adjoined ϵ); analysis proofs that are very difficult for humans and machines are often easy to prove in non-standard analysis.

Lightweight Probability Theory for Verification

Joe Hurd[*]

Abstract. *There are many algorithms that make use of probabilistic choice, but a lack of tools available to specify and verify their operation. The primary contribution of this paper is a lightweight modelling of such algorithms in higher-order logic, together with some key properties that enable verification. The theory is applied to a uniform random number generator and some basic properties are established. As a secondary contribution, all the theory developed has been mechanized in the* `hol98` *theorem-prover.*

1 Summary

There are many algorithms with a probabilistic specification, and more that make use of probabilistic choice in their operation. A prominent example from mathematics is the Miller-Rabin primality test, which takes a number n and returns either PRIME or COMPOSITE. If n actually is prime then it is guaranteed to return PRIME, and if n is composite then it will return COMPOSITE with probability at least one half. Successive calls are independent, so if n is composite then s consecutive results of PRIME will occur with probability at most 2^{-s}.

Our goal is to verify in a higher-order logic theorem-prover that such a probabilistic program fulfills its specification. This requires creating two new logical theories: the first similar to a programming language in which to express the program; and the second a formalization of probability theory in which to express the specification.

The programming language we use is the language of higher-order logic functions, augmented so that a probabilistic function of type $\alpha \to \beta$ is

---
[*]Supported by an EPSRC studentship.

modelled by a deterministic function of type $\alpha \rightarrow B^\infty \rightarrow \beta \times B^\infty$. Elements of type B^∞ are infinite sequences of booleans (modelling a random number generator), and functions are expected to use booleans from the head of the sequence and pass back a sequence of 'unused' booleans with the usual result, just as if the sequence was a global random boolean generator. This method of modelling probabilistic functions can be described very neatly using state transforming monads, and this is the approach taken in pure functional languages such as Haskell.

Formalizing the concepts of *probability* and *independence* in mathematics was a challenging task pioneered by Kolmogorov in the 1930s, and resulted in the mathematical theory of measurable sets. In this work we build upon Harrison's construction of the real numbers in higher-order logic, and formalize enough measure theory to define the probability of any 'event' produced by a (terminating) probabilistic program written in the above programming language.

Our results so far are:

- an implementation of a uniform random number generator, guaranteed to produce numbers with a particular probability distribution;

- a set of monadic primitives for constructing probabilistic programs, with proofs that all programs constructed using these primitives (including Haskell probabilistic programs) have strong independence properties;

- support for executing probabilistic programs in the logic (using a pseudo-random boolean generator), for purposes of debugging and also the proving of certain properties (if Miller-Rabin returns COMPOSITE for an input n, then $\neg$(prime n) is a theorem).

For future work, we note there are computer packages (such as Mathematica[1] and SPSS[2]) that offer a range of statistical analysis functions, but so far there is no theorem-proving system able to verify their results. This interaction has benefitted both theorem-provers and computer algebra systems in other domains; perhaps we could apply the technology to probability and statistics.

---

[1] http://www.wolfram.com/
[2] http://www.spss.com/

St Andrews CAAR Group:
Poster Abstract

Tom Kelsey

The St Andrews Computer Algebra and Automated Reasoning group is working in three areas related to computation and proof for real numbers and real-valued functions:

1. The implementation of exact real numbers arithmetic in Aldor, the extension language of the AXIOM computer algebra system. Exact real arithmetic is computation over infinite sequences of data, using a lazy data type with a procedure for obtaining the nth finite data item. Our project aims to augment the numeric and symbolic tools used by the working mathematician with a library of exact routines. We currently have a prototype Aldor library in which real numbers in the interval $[-1, 1]$ are represented by streams of elements from $GF3$, the field containing exactly 3 elements. This representation is computable with respect to the type-2 theory of effectivity, and utilises features of the Aldor basicmath library.

2. The design and implementation of a real-valued function toolkit for the theorem prover PVS. Our aim is to develop theorem proving support for computer algebra systems. The toolkit focuses on the formal development of transcendental functions in PVS, and the checking of continuity of functions. We have been able to prove the continuity of functions such as

$$exp(x^2 + |1 - x|), \quad exp(cos(x) + 1), \quad and \quad \frac{1}{|x| + 1} + \frac{2}{|x| + 2}.$$

The toolkit has been applied Maple solutions of partial differential equations by providing (i) a verification of the solution, and (ii) a list of the implicit constraints upon which the solution relies.

3. The construction of a verifiable symbolic definite integral table lookup (VSDITLU): a system which matches a query, comprising a definite integral with parameters and side conditions, against an entry in

a verifiable table and uses a call to a library of facts about the reals in the theorem prover PVS to aid in the transformation of the table entry into an answer. Our system is able to obtain correct answers in cases where standard techniques implemented in computer algebra systems fail. We have presented a full model of such a system, and described a a prototype implementation showing the efficacy of such a system: for example, the prototype is able to obtain correct answers in cases where computer algebra systems [CAS] do not. We have extended Fateman's web-based table by including parametric limits of integration and queries with side conditions.

OpenXM — an Open System to Integrate Mathematical Software

Masahide Maekawa Yukio Okutani
Nobuki Takayama Yasushi Tamura Masayuki Noro
Katsuyoshi Ohara

OpenXM (Open message eXchange protocol for Mathematics) is a project aiming to integrate data, control and user interfaces with the following fundamental architecture.

1. Communication is an exchange of *OX (OpenXM) messages*, which are classified into three types: DATA, COMMAND, and SPECIAL. *OX data messages* wrap mathematical data. We use standards of mathematical data formats such as OpenMath and MP as well as our own data format *CMO (Common Mathematical Object format)*.

2. Each server, which provides services to other processes, is a stack machine. The stack machine is called the *OX stack machine*. Existing mathematical software tools are wrapped with this stack machine.

3. Any server may have a hybrid interface; it may accept and execute not only stack machine commands, but also its original command sequences. For example, if we send the following string to the `ox_asir` server (OpenXM server based on Risa/Asir)
   ```
   "fctr(x^100-y^100);"
   ```
   and call the stack machine command
   ```
   SM_executeStringByLocalParser
   ```
   then the server executes the asir command `fctr(x^100-y^100);` (factorize $x^{100} - y^{100}$ over $\mathbf{Q}$) and pushes the result onto the stack.

4. Network transparent supports for controlling servers are provided. For example OpenXM defines a robust reset procedure to restart computations without any confusion in I/O buffers. It is very useful for debugging programs running on distributed environment.

OpenXM is supposed to be used to connect mathematical softwares in the academic community and to test new ideas in distributed algorithms in mathematics. Currently the following servers are available in the OpenXM package.

ox_asir A server for Risa/Asir, a general-purpose computer algebra system. It provides almost all functionalities of Risa/Asir such as polynomial factorization, Gröbner basis computation and primary ideal decomposition.

ox_sm1 A server for Kan/sm1, a system for computation in the ring of differential operators including computation of Gröbner bases and cohomology groups.

ox_phc A server for PHC pack, a general-purpose solver for polynomial systems by homotopy continuation.

ox_tigers A server for TiGERS, a system to enumerate all Gröbner bases of affine toric ideals. It can be used to determine the state polytope of a given affine toric ideal.

ox_gnuplot A server for GNUPLOT, a famous plotting tool.

ox_math A server for Mathematica.

OMproxy A server for translation between CMO and OpenMath/XML expressions. It is written in Java. This module provides Java classes OXmessage, CMO, and SM for the OpenXM protocol, too.

In addition to these servers, Risa/Asir, Kan/sm1 and Mathematica can act as clients. For example, the following is a command sequence to ask $1 + 1$ from the Asir client to the ox_sm1 server:

```
P = sm1_start(); ox_push_cmo(P,1); ox_push_cmo(P,1);
ox_execute_string(P,"add"); ox_pop_cmo(P);
```

The OpenXM package is currently implemented on TCP/IP, which uses the client-server model. The OpenXM on MPI is currently tested on Risa/Asir, where collective operations such as broadcast and reduction are implemented to achieve a real speedup for massively parallel computation. C library interfaces are available for some servers. The source code and binary packages are distributed from http://www.openxm.org. The current version is 1.1.2. Documents and a full paper including references are available from this site, too.

Presentation of the Foc Project

Renaud Rioboo

Abstract. *The purpose of the Foc project which started at fall 1997 is to provide a framework for certified computer algebra. Its typical user is a computer algebra developer who is concerned with program correction. It's goal is to provide a set of tools to produce effective code as a combination of Ocaml classes and modules and effective proofs as Coq's theorems. It should also provide concrete syntaxes designed to help the programmer to give the proof together with the algorithm.*

1 The FOC Framework

The Foc ([1]) model is akin to the Axiom's ([3], [5]) model and defines *entities* which have a *representation*. Functionalities available over entities are then described by *species* which are "frozen" into *collections*. Roughly speaking entities can be assimilated to Axiom's categories and collections to Axiom's domains.

The main difference between the Foc model and the Axiom's model is that we do not redefine a language and do not abstract concrete data types. These are external to the mathematical view of the library and can receive different Ocaml ([4]) or Coq ([2]) instantiations in order to respectively run or prove properties of the functionality that we implement. For instance when implementing polynomials as lists of couples we will use properties of lists in Coq or use Ocaml list operations in order to prove and implement a library of univariate polynomials.

In particular the run time system is statically typed by Ocaml's type system and for a given collection the representation of its entities is always passed as type argument of an Ocaml class. This type is then abstracted in an Ocaml module to provide type safety.

2 Description of the Library

The current FOC library is made of around a hundred species which specify common algebraic structures. In the Foc model collections are created by the programmer by freezing species which specify a reprensentation together with functionalities. This is achieved by creating an Ocaml object together with hiding the concrete representation inside an Ocaml module.

Currently the library implements basic small integers as Ocaml `int`. These entities are then viewed as an additive monoid (which are mainly used for degrees of polynomials), a commutative ring or a field for computation modulo small numbers. Arbitrary large integers are represented as either Ocaml standard `big_int` or using an interface with the `Gmp` library. These are then viewed as an euclidean ring by two species.

The library then provides functionalities to view basic Ocaml products and basic Ocaml vectors as abstract sets, monoids, rings. The current library implement basic arithmetics for distributed polynomials, sparse univariate polynomials are then a special instance of distributed polynomials.

Recursive polynomials are implemented as an iteration of the distributed case allowing the programmer to mix recursive and distributed representations. This slightly more general than what is implemented for instance in Axiom or in the BasicMath Aldor's library.

We will give some points of comparisons with Axiom which show that small computations are somewhat faster in Foc than in Axiom. Whereas big computations are much faster using Foc together with the `big_int` Ocaml library, these are even faster using the `Gmp` library.

References

[1] S. Boulmé, T. Hardin, D. Hirschkoff, V. Ménissier-Morain, and R. Rioboo. *On the way to certify Computer Algebra Systems* Proceedings of the 1999 Calculemus Workshop of FLOC'99, 1999

[2] Coq project, *The Coq Proof Assistant Reference Manual*, 1999.

[3] D. Jenks Richard and Robert S. Stutor. *AXIOM, The Scientific Computation System*. Springer-Verlag, 1992.

[4] Xavier Leroy. *The Objective Caml system, release 2.0*. Software and documentation available: http://caml.inria.fr/ocaml/, 1998.

[5] S. Watt, P. Broadbery, S. Dooley, P. Iglio, S. Morrison, J. Steinbach, et R. Sutor. *AXIOM Library Compiler User Guide*. NAG Ltd, March 1995.

Author Index

Authors' Affiliations

Andrew A. Adams
    School of Computer Science,
    University of St Andrews, Scotland
    aaa@cs.st-and.ac.uk

Noriko H. Arai
    Hiroshima City University, Hiroshima, Japan
    narai@cs.hiroshima-cu.ac.jp

Alessandro Armando
    DIST, Università di Genova, Italy
    armando@dist.unige.it

Clemens Ballarin
    Fakultät für Informatik,
    Universität Karlsruhe, Germany
    ballarin@ira.uka.de

Grzegorz Bancerek
    Institute of Computer Science,
    University of Białystok, Poland
    bancerek@mizar.org

Henk Barendregt
    Mathematics and Computer Science,
    University of Nijmegen, Nijmegen, The Netherlands
    henk@cs.kun.nl

Christian Bauer
   Institute of Physics,
   Johannes Gutenberg University, Mainz, Germany
   Christian.Bauer@uni-mainz.de

Marco Benedetti
   Dipartimento di Informatica e Sistemistica,
   Università di Roma "La Sapienza", Rome, Italy
   mabe@dis.uniroma1.it

Christoph Benzmüller
   School of Computer Science,
   The University of Birmingham, Birmingham, England
   C.E.Benzmuller@cs.bham.ac.uk

Bruno Buchberger
   Research Institute for Symbolic Computation,
   Johannes Kepler University, Schloß Hagenberg, Linz, Austria
   Bruno.Buchberger@risc.uni-linz.ac.at

Alan Bundy
   Division of Informatics,
   University of Edinburgh, Edinburgh, Scotland
   bundy@dai.ed.ac.uk

Jacques Calmet
   Fakultät für Informatik,
   Universität Karlsruhe, Germany
   calmet@ira.uka.de

Olga Caprotti
   Mathematics and Computing Science,
   Eindhoven University of Technology, Eindhoven, The Netherlands
   olga@win.tue.nl

Arjeh Cohen
   Mathematics and Computing Science,
   Eindhoven University of Technology, Eindhoven, The Netherlands
   amc@win.tue.nl

Simon Colton
  Division of Informatics,
  University of Edinburgh, Edinburgh, Scotland
  simonco@dai.ed.ac.uk

Anatoli I. Degtyarev
  Department of Computing and Mathematics,
  Manchester Metropolitan University, Manchester, England
  A.Degtyarev@cs.man.ac.uk

Claudio Dupré
  Research Institute for Symbolic Computation,
  Johannes Kepler University, Schloß Hagenberg, Linz, Austria
  cdupre@risc.uni-linz.ac.at

Jacques Fleuriot
  Division of Informatics,
  University of Edinburgh, Edinburgh, Scotland
  jdf@dai.ed.ac.uk

Alexander Frink
  Institute of Physics,
  Johannes Gutenberg University, Mainz, Germany
  Alexander.Frink@uni-mainz.de

Fausto Giunchiglia
  DISA, Università di Trento, ITC-IRST, Trento, Italy
  fausto@irst.itc.it

Gaston H. Gonnet
  Institute for Scientific Computation,
  ETH Zürich, Zürich, Switzerland
  gonnet@inf.ethz.ch

Gert-Martin Greuel
  Fachbereich Mathematik,
  Universität Kaiserslautern, Kaiserslautern, Germany
  greuel@mathematik.uni-kl.de

Alex Heneveld
    Division of Informatics,
    University of Edinburgh, Edinburgh, Scotland
    heneveld@cogsci.ed.ac.uk

Joe Hurd
    Computer Laboratory,
    University of Cambridge, Cambridge, England
    joe.hurd@cl.cam.ac.uk

Mateja Jamnik
    School of Computer Science,
    The University of Birmingham, Birmingham, England
    M.Jamnik@cs.bham.ac.uk

Tudor Jebelean
    Research Institute for Symbolic Computation,
    Johannes Kepler University, Schloß Hagenberg, Linz, Austria
    Tudor.Jebelean@risc.uni-linz.ac.at

Tom Kelsey
    School of Computer Science,
    University of St Andrews, St Andrews, Scotland
    tom@dcs.st-and.ac.uk

Manfred Kerber
    School of Computer Science,
    The University of Birmingham, Birmingham, England
    M.Kerber@cs.bham.ac.uk

Michael Kohlhase
    Fachbereich Informatik,
    Universität des Saarlandes, Saarbrücken, Germany
    kohlhase@ags.uni-sb.de

Boris Konev
    Steklov Institute of Mathematics, St. Petersburg, Russia
    konev@pdmi.ras.ru

Richard Kreckel
 Institute of Physics,
 Johannes Gutenberg University, Mainz, Germany
 Richard.Kreckel@uni-mainz.de

Franz Kriftner
 Research Institute for Symbolic Computation,
 Johannes Kepler University, Schloß Hagenberg, Linz, Austria
 Franz.Kriftner@risc.uni-linz.ac.at

Alexander V. Lyaletski
 Cybernetics Department,
 Taras Shevchenko University of Kyiv, Kiev, Ukraine
 lav@tk.cyb.univ.kiev.ua

Ewen Maclean
 Division of Informatics,
 University of Edinburgh, Edinburgh, Scotland
 ewenm@dai.ed.ac.uk

Masahide Maekawa
 Department of Mathematics,
 Kobe University, Kobe, Japan
 maekawa@math.sci.kobe-u.ac.jp

Ryuji Masukawa
 Hiroshima City University,
 Hiroshima, Japan
 masukawa@log05.logic.cs.hiroshima-cu.ac.jp

Andreas Meier
 Fachbereich Informatik,
 Universität des Saarlandes, Saarbrücken, Germany
 ameier@ags.uni-sb.de

Marina K. Morokhovets
 Digital Automata Theory Department,
 Glushkov Institute of Cybernetics, Kiev, Ukraine
 mmk@d105.icyb.kiev.ua

Koji Nakagawa
    Research Institute for Symbolic Computation,
    Johannes Kepler University, Schloß Hagenberg, Linz, Austria
    Koji.Nakagawa@risc.uni-linz.ac.at

Masayuki Noro
    Fujitsu Laboratories,
    Kawasaki, Japan
    noro@para.flab.fujitsu.co.jp

Katsuyoshi Ohara
    Department of Computational Science,
    Kanazawa University, Kanazawa, Japan
    ohara@kappa.s.kanazawa-u.ac.jp

Yukio Okutani
    Department of Mathematics,
    Kobe University, Kobe, Japan.

Martijn Oostdijk
    Mathematics and Computing Science,
    Eindhoven University of Technology
    martijno@win.tue.nl

Gerhard Pfister
    Fachbereich Mathematik,
    Universität Kaiserslautern, Kaiserslautern, Germany
    pfister@mathematik.uni-kl.de

Silvio Ranise
    DIST, Università di Genova, Italy
    silvio@dist.unige.it

Renaud Rioboo
    Laboratoire d'Informatique de Paris 6,
    Université Pierre et Marie Curie, Paris, France
    Renaud.Rioboo@lip6.fr

Piotr Rudnicki
Department of Computing Science,
University of Alberta, Edmonton, Alberta, Canada
piotr@cs.ualberta.ca

Hans Schönemann
Fachbereich Mathematik,
Universität Kaiserslautern, Kaiserslautern, Germany
hannes@mathematik.uni-kl.de

Christoph Schwarzweller
Wilhelm-Schickard-Institute for Computer Science,
University of Tübingen, Tübingen, Germany
schwarzw@informatik.uni-tuebingen.de

Roberto Sebastiani
DISA, Università di Trento, Trento, Italy
rseba@cs.unitn.it

Alan Smaill
Division of Informatics,
University of Edinburgh, Edinburgh, Scotland
smaill@dai.ed.ac.uk

Volker Sorge
Fachbereich Informatik,
Universität des Saarlandes, Saarbrücken, Germany
sorge@ags.uni-sb.de

Nobuki Takayama
Department of Mathematics,
Kobe University, Kobe, Japan
taka@math.s.kobe-u.ac.jp

Yasushi Tamura
Department of Mathematics,
Kobe University, Kobe, Japan.

Simon Thompson
Computing Laboratory,
University of Kent, Canterbury, England
S.J.Thompson@ukc.ac.uk

Paolo Traverso
ITC-IRST, Trento, Italy
leaf@irst.itc.it

Andrzej Trybulec
Institute of Mathematics,
University of Białystok, Poland
trybulec@math.uwb.edu.pl

Daniela Văsaru
Research Institute for Symbolic Computation,
Johannes Kepler University, Schloß Hagenberg, Linz, Austria
dvasaru@risc.uni-linz.ac.at

Wolfgang Windsteiger
Research Institute for Symbolic Computation,
Johannes Kepler University, Schloß Hagenberg, Linz, Austria
Wolfgang.Windsteiger@risc.uni-linz.ac.at

Daniele Zini
DIST, Università di Genova, Italy
danielez@dist.unige.it

T - #0037 - 160425 - C0 - 229/152/15 [17] - CB - 9781568811451 - Gloss Lamination